BESTSELLING
BOOK SERIES

# Evolution For Dum...

## Scientific Process at a Glance

The thing that makes a science a science is the adherence to the scientific process:

1. **Make observations about the natural world.**

2. **Formulate a hypothesis.**

   The hypothesis serves as the scientist's starting point; maybe it's right, and maybe it's wrong. They key is to do enough testing to find out.

3. **Gather additional data to test this hypothesis.**

   As your data accumulates, it either supports your hypothesis, or it forces you to revise or abandon the hypothesis.

   *Remember:* The hypothesis scientists come up with must be *falsifiable*. That is, scientists must be able to imagine some set of results that would cause them to reject the theory, and then they must test those ideas out.

4. **Continue testing (if the data from Step 3 supports your hypothesis) or revise your hypothesis and test again.**

   After an overwhelming amount of information accumulates in support of the hypothesis, you elevate the hypothesis to a theory.

5. **If, at anytime in the future, new data arises that causes you to revise or reject your theory, then you revise or reject it and start again at Step 1.**

   Real scientists *never* ignore facts or observations in order to protect a hypothesis or theory, even one that they're particularly fond of.

## The Hominid Family Tree

Your family tree holds more than Uncle Joe and Great-grandma Myrtle. Take a look these distant and not-so-distant relations:

| Species* | Where Found | Lived (mya= million years ago) | Interesting Characteristics |
|---|---|---|---|
| A. anamensis | Kenya | 4.2 to 3.9 mya | Probably walked upright |
| A. afarensis | Ethiopia, Kenya | 3.6 to 2.9 mya | Walked upright. Most famous member (to us anyway) is Lucy, the nearly complete fossil found in 1974 |
| A. africanus | South Africa | 3 to 2 mya | Teeth more human-like than ape-like, probably bipedal |
| A. aethiopicus | Ethiopia | 2.7 to 2.3 mya | Considered a transitional species between A. afarensis and A. boisei |
| A. garhi | Ethiopia | 2.5 mya | Possibly the earliest tool user |
| A. boisei | Tanzania, Ethiopia | 2.3 to 1.4 mya | Formerly considered to be a direct human ancestor until H. habilis was discovered |
| A. robustus | South Africa | 1.8 to 1.5 mya | May have used tools to dig up edible roots |
| H. rudolfensis | Kenya, Tanzania | 2.4 to 1.8 mya | Bipedal with a large brain |
| H. habilis | Kenya | 2.3 to 1.6 mya | "Handy man"; used tools, brain larger and more human-shaped, possibly capable of rudimentary speech |
| H. ergaster | Eastern and Southern Africa | 1.9 to 1.4 mya | Made some nice tools, had smaller teeth |
| H. erectus | Republic of Georgia, Kenya, China, Indonesia, Europe | 1.9 to 0.3 mya (and possibly 50,000 years ago) | Definitely used tools, probably discovered fire, and may have lived at same time as modern humans |
| H. heielbergensis | Africa, Europe | 600,000 to 100,000 years ago | Brain size equal to modern humans, found with tools sharp enough to slice through animal hides, almost certainly used fire |
| H. neanderthalensis | Europe, Middle East | 250,000 to 30,000 years ago | Lived mostly in cold climates, shared the earth with H. sapiens, may have had a complex social system that included care for the elderly and burial rituals |
| H. sapiens | Worldwide | 100,000 years ago to today | Large brains (not always used) and ability to manipulate tools, situations, and the emotions of other H. sapiens |

*A = Australopithecus, H = Homo

# Evolution For Dummies®

Cheat Sheet

## Interpreting Evolution Articles

Ever found yourself scratching your head at some of the terms used in the science section of the newspaper or a science article in a magazine? These definitions might come in handy. Keep them next to your morning coffee.

**Adaptations:** Changes resulting from natural selection.

**Allele (plural _alleles_):** The specific DNA sequence found at a given locus in an individual. A haploid individual has one allele at every locus while a diploid individual has two alleles at each locus (one on each set of chromosomes), which can be the same or different.

**Artificial selection:** The process of selection when people control which characters are favored—for example selectively breeding cows that make the most milk to produce the next generation of dairy cows.

**Chromosome:** The cellular structures that contain DNA. Humans, a diploid organism, have 23 pairs of chromosomes.

**Diploid genome:** The genome of an organism that has two sets of chromosomes. In sexually reproducing organisms, diploid parents each contribute one set of chromosomes to the offspring, producing a new diploid individual whose genome is a combination of some of the DNA from each parent. Examples of diploid organisms include mammals, birds, many plants, and so on.

**DNA (deoxyribonucleic acid):** A long molecule made up of four different subunits (or nucleotides, which you can think of as a four-letter alphabet). The sequence of the four different nucleotides govern the specific details of traits. While almost all organisms have DNA as the genetic material, a few (some viruses) use a slightly different molecule (RNA, ribonucleic acid) but the process is otherwise the same.

**DNA sequence:** The exact arrangement of the four nucleotides in a specific individual. The sequence information can be for the entire genome or just some location of interest.

**Evolution:** A change in the percentage of inherited (heritable) traits in a group of organisms over time. For evolution, time is measured in generations (which is one of the reasons that bacteria evolve faster than elephants).

**Evolutionary theory:** The field of scientific investigation that works to understand what processes are responsible for the evolutionary changes we observe and what the consequences of those changes are.

**Fitness:** A measure of an organism's ability to contribute offspring to the next generation.

**Gene:** The classic unit of heredity that governs the traits that are passed from parent to offspring. The term predates an understanding of how the process of heredity actually works, which involves DNA. Therefore, in science articles, _gene_ primarily serves as an easy-to-understand, if not exactly precise, stand in for _locus_ and _allele_, which more precisely identify the exact units of heredity.

**Genetic drift:** Random factors—volcanoes erupting, trees falling, or airplanes crashing, for example—that impact the gene frequency in subsequent populations.

**Genome:** The sum total of all of an organism's DNA.

**Genotype:** The specific combination of alleles that an individual organism has. Genotype does not map directly to phenotype (or physical traits) because of the effect of environmental factors.

**Haploid genome:** The genome of an organism with a single set of chromosomes. Examples of haploid organisms include bacteria and fungi which produce asexually (new individuals simply divide from existing ones). _**Note:**_ Diploid individuals produce haploid gametes (sperm and egg).

**Locus (plural _loci_):** A particular location in an organism's genome where the information for a particular trait resides. The eye color locus, for example, is the place in an individual organism's genome that has the DNA sequence controlling eye color.

**Mutations:** Changes in the DNA sequence caused by errors in DNA replication or such factors (like radiation) that can cause DNA damage.

**Natural selection:** The process of selection when the natural environment is the selective force.

**Neutral evolution:** Evolution as the result of genetic drift. When two different alleles are selectively neutral—that is, they don't differ in fitness—changes in their relative frequencies can only be caused by random events.

**Phenotype:** The physical traits that the organism has, including things like body structure, wing span, running speed, and so on. Phenotype is a product of both the genotype and the effects of the environment.

**Selection:** When a particular character is favored such that organisms that possess that character are more likely to contribute offspring to the next generation. If the character under selection is heritable, then the frequency of that character in future generations increases. Selection acts on phenotypes rather than genotypes.

**Speciation:** When a group of individuals in a species evolves differently from the rest of the species, leading to the accumulation of enough genetic differences to prevent the two groups from interbreeding.

## For Dummies: Bestselling Book Series for Beginners

# Evolution

## FOR
## DUMMIES®

**by Greg Krukonis, PhD, and Tracy Barr**

**WILEY**

Wiley Publishing, Inc.

**Evolution For Dummies**®

Published by
**Wiley Publishing, Inc.**
111 River St.
Hoboken, NJ 07030-5774
www.wiley.com

Copyright © 2008 by Wiley Publishing, Inc., Indianapolis, Indiana

Published simultaneously in Canada

For general information on our other products and services, please contact our Customer Care Department within the U.S. at 800-762-2974, outside the U.S. at 317-572-3993, or fax 317-572-4002.

For technical support, please visit www.wiley.com/techsupport.

Wiley also publishes its books in a variety of electronic formats. Some content that appears in print may not be available in electronic books.

Library of Congress Control Number: 2008922285

ISBN: 978-0-470-11773-6

Manufactured in the United States of America

10  9  8  7  6  5  4  3  2  1

WILEY

# About the Authors

**Dr. Greg Krukonis:** Greg Kukonis has a Bachelor of Arts degree in Biology from the University of Pennsylvania and a PhD from the University of Arizona, Department of Ecology and Evolutionary Biology. He has been a postdoctoral researcher at Wesleyan University in Middletown, Connecticut, and Stanford University. He is currently an adjunct assistant professor of biology at Lewis and Clark College in Portland, Oregon.

**Tracy Barr:** Tracy Barr is a professional writer and editor who has authored or co-authored several other books for Wiley, including *Adoption For Dummies, Cast-Iron Cooking For Dummies, Yorkshire Terriers For Dummies,* and *Latin For Dummies.* She lives in Indianapolis with her husband and four children.

# Dedication

**From Greg:** To my family, to Tarsah, and to the mentors, colleagues, and students who have shared their insights, their enthusiasm, and their friendship. And to everyone who's ever wondered what evolution really is and to anyone who's ever been struck by the beautiful and amazing diversity of life on Earth.

# Acknowledgments

**From Greg:** I'd like to thank my friends and colleagues who were always telling me, "Greg, you should write more" and who encouraged me throughout the project. It turns out that I did have a book in me, but it definitely took a village to help me find my voice. To say that I got by with a little help from my friends doesn't begin to describe my gratitude for the both the general encouragement and the numerous specific helpful suggestions I was so fortunate to receive.

I would also like to give thanks for the limitless amount of patience that has been shown me in the face of an ever-changing and over extended schedule, late nights, canceled plans, and the various unexpected challenges inherent in such a project. A special thanks goes to Stacy Kennedy and the other folks at Wiley who were instrumental in bringing this book to life.

Finally I'd like to acknowledge the many conversations about evolution I've had with random strangers I've met on airplanes, in coffee shops, and at cocktail parties — everywhere from the top of a cold mountain in New Hampshire to a toasty warm pub in New Zealand. I learned a lot from those conversations about what people want to know about evolution and where things get confusing, and I've tried to address these areas in this book.

## Publisher's Acknowledgments

We're proud of this book; please send us your comments through our Dummies online registration form located at www.dummies.com/register/.

Some of the people who helped bring this book to market include the following:

### Acquisitions, Editorial, and Media Development

**Acquisitions Editor:** Stacy Kennedy

**Copy Editor:** Kathy Simpson

**Technical Editor:** Veronique Delesalle, Professor of Biology, Gettysburg College

**Senior Editorial Manager:** Jennifer Ehrlich

**Editorial Supervisor and Reprint Editor:** Carmen Krikorian

**Editorial Assistants:** Erin Calligan Mooney, Joe Niesen, Leeann Harney

**Art Coordinator:** Alicia South

**Cover Photos:** © Sally A. Morgan; Ecoscene/CORBIS

**Cartoons:** Rich Tennant (www.the5thwave.com)

### Composition Services

**Project Coordinator:** Erin Smith

**Layout and Graphics:** Claudia Bell, Alissa D. Ellet, Joyce Haughey

**Proofreaders:** John Greenough, C.M. Jones

**Indexer:** Glassman Indexing Services

---

**Publishing and Editorial for Consumer Dummies**

**Diane Graves Steele,** Vice President and Publisher, Consumer Dummies

**Joyce Pepple,** Acquisitions Director, Consumer Dummies

**Kristin A. Cocks,** Product Development Director, Consumer Dummies

**Michael Spring,** Vice President and Publisher, Travel

**Kelly Regan,** Editorial Director, Travel

**Publishing for Technology Dummies**

**Andy Cummings,** Vice President and Publisher, Dummies Technology/General User

**Composition Services**

**Gerry Fahey,** Vice President of Production Services

**Debbie Stailey,** Director of Composition Services

# Contents at a Glance

# Table of Contents

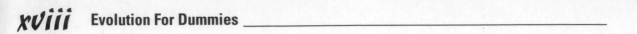

# Introduction

. . . . . . . . . . . . . . . . . . . . . . . . . . . . . . . . . . . . . . . . . . . . . . .

*E*volution is the process by which populations and species change over time. The principles of evolution explain why life on Earth is so varied and why organisms are the way they are. The study of evolution is not only interesting for its own sake, but it's also a fundamental part of the biological sciences. You can't understand (or combat) disease, can't understand the history of species (or the world, for that matter) — can't do a lot of things, in fact, without understanding evolution.

Simply put, evolution is *the* key scientific principle behind every substantive thing we know about biology, the study of living things. And its main points are remarkably easy to understand.

So why did I write a whole book about evolution? Because a lot of people are confused about exactly what evolution is, what it does, how it works, and why it's important. This book helps you sort everything out.

## About This Book

You may have the sense that only the super-smart can understand any branch of science. If you didn't see the point of being able to identify the parts of a cell, or you didn't like memorizing the periodic table of elements, your experience confirms that sense. And you've probably figured out that you're no Einstein, but — here's a secret — most scientists (including yours truly) aren't Einsteins either.

In fact, the smart money says that Einstein was so smart that most of the rest of us aren't smart enough even to know how smart he was. A possible exception may be someone like Stephen Hawking, but none of us is smart enough to know how smart *he* is, either. But I digress. The point is that you don't have to be an Einstein or a Hawking to "get" science. As I'm fond of saying to my students, evolution isn't rocket science — and for that matter, rocket science isn't rocket science either.

I wrote this book to help you overcome whatever natural reluctance you may have about reading an evolution book and to clear away the confusion caused by all the bad info out there. To that end, I've divided each chapter into sections that contain information about some component of evolution or one of the many hot topics that evolutionary biology helps people understand, such as:

- ✔ What natural selection is and how it works
- ✔ How to trace the evolutionary history of organisms
- ✔ The evolutionary component of social systems
- ✔ Where modern man came from
- ✔ How diseases evolve, and what scientists are doing to fight them

If there's one thing I want you to take away from this book, it's this: The lion's share of science, if explained clearly, is accessible to everyone. Sure, you have to be an expert in the field to fully grasp the importance of the details. But the broad strokes should be accessible to everyone, and that is certainly the case for evolutionary biology.

# Conventions Used in This Book

To help you navigate easily, this book uses a few standard conventions:

- ✔ *Italic* is used for emphasis and to highlight new words or terms that are defined.
- ✔ `Monofont` is used for Web addresses.
- ✔ You'll also see quite a bit of *we* in this book. Sometimes, *we* refers to me and other experts in the field of evolution. At other times, *we* refers to me and you. Just like you, I am constantly amazed by and in awe of the beauty of evolution.

# What You're Not to Read

I love everything about evolution: the big points, the little points, the so-esoteric-that-no-one-but-other-evolutionary-biologists-will-find-them-even-remotely-interesting points. I'd love to think that you're just as enamored of evolution as I am, but being a realist (and scientist), I have to face facts: You probably aren't. So to meet my need (to include as much information as possible) and yours (to get to the key points quickly), I've made it easy for you to identify material that you can safely skip:

✔ **Text in sidebars:** The sidebars are the shaded boxes that appear here and there. They aren't necessary reading.

✔ **Anything with a Technical Stuff icon attached:** This information is interesting but not critical to your understanding of evolution.

# Foolish Assumptions

Every book is written with a particular reader in mind, and this one is no different. As I wrote this book, I made a few assumptions about you:

✔ You've heard about Charles Darwin but aren't quite clear about what he actually said or why it was so revolutionary.

✔ You're confused by all the contradictory claims you hear about evolution and want to know what the science actually says.

✔ You're curious about the evolution of species, both in general — where do they come from, for example — and more close to home, such as the evolution of our own species and the diseases that plague us and which seem to grow more dangerous with every new generation.

✔ You've seen the 1960 film *Inherit the Wind,* and beyond noting that Darrin Stephens is the defendant, you want to know the science behind the events depicted.

✔ Even though you know that 99.9999 percent of scientists accept the theory of evolution, you need proof that these 99.9999 percent aren't wrong.

# How This Book Is Organized

To help you find information that you're looking for, this book is divided into five parts. Each part covers a particular aspect of evolution and contains chapters relating to that part.

## Part 1: What Evolution Is

Look up the word *evolution* in a dictionary, and you'll come across a definition that says something about change or maybe change through time. That's good as far as it goes. But in the context of biology, evolution refers to specific changes — genetic changes — in a group of organisms through time. That concept isn't so hard to grasp, but you may be surprised by how revolutionary the idea of evolution was in the mid-1800s, when Darwin came up with his theory explaining what could cause such changes (natural selection).

Back then, the concept that species could change over time — even the concept of vast time spans — was foreign and frightening to most people. But facts are facts, evidence has a way of piling up, and the science of evolutionary biology has progressed in the century and a half since Darwin's major insight.

This part introduces the key principles of evolution by natural selection. And because to grasp the main idea, you need to know a bit about genetics, the part includes a brief discussion of that topic, too. If it makes you feel any better (and it should), reading this short discussion of genetics puts you in the position of knowing more about genetics and heredity than Darwin himself did.

## Part II: How Evolution Works

Sometimes, evolution is the result of natural selection. Other times, it's the result of random factors (genetic drift). Populations have variability; not all the individuals are the same, and sometimes individuals with particular genetic traits leave more descendant than others. That's evolution in a nut shell: The next generation is genetically different from the last one because not everybody's genes made it! These changes can have big effects on populations. Sometimes they end up with altered proportions of different variants (more fast cheetahs than slow ones, for example). Sometimes, they lose genetic variation, and sometimes, just sometimes, populations speciate (that is, form a new species).

You can consider this part to be the nuts-and-bolts section of the book, because it explains that biological variation exists, where this variation comes from, and the different ways it can change through time. Plus this is the part where I explain how scientists can watch evolution happen both in laboratory experiments and in nature, as well as how they can use data about species today to come up with strong hypothesis about evolution in the past.

## Part III: What Evolution Does

Evolution is no more complicated than genetic changes accumulating through time. Sounds almost boring, yet it's anything *but* boring. Because all those changes in the DNA, which you can't even see (outside a biology lab) influence all the things about living creatures that you cannot only see, but also be amazed by. Look out the window at nature's diversity: Evolution did that! Evolution has a pretty big impact on lots of things you can observe about life, such as:

✔ Physical characteristics (petal color, length of tail, eye color, and so on)

✔ Body shape (number of fins, fingers, limbs, and heads, for example)

✔ Sexual selection (who mates with whom, how, and why)

✔ Life histories (reproduction and life spans)

✔ Social behaviors (competitive, altruistic, and so on)

This part covers 'em all.

# Part IV: Evolution and Your World

Two things hold folks' attention better than anything else: themselves and things that affect them. This part covers both topics, beginning with human evolution to explain where we came from (out of Africa), whether we're unique among all the animals in creation (it turns out that we aren't; quite a few other hominid species preceded us, and a couple even shared the Earth with us for a while), and how we continue to evolve.

The remainder of the part delves into antibiotic resistance and the evolution of two scourges: HIV and influenza. Why the shift from the exalted Us to the microbial Them? Because these buggers can and do wreak havoc on humans by evolving so quickly and in response to the very medications we use to fight them. Perhaps you've seen on the news that bacteria have "acquired" or "developed" antibiotic resistance. Those are just other ways of saying that these bacteria have evolved resistance to our antibiotics — a problem that we need to stay on top of.

# Part V: The Part of Tens

Throughout the book, I spend a lot of time talking about the fossil record and adaptations, explaining what they are and why they're important to evolutionary study. But in this part, I list the fossils and adaptations that are particularly fun or revealing.

I also include the only response you're going to find to the challenges people throw at evolution. The purpose of these challenges isn't to clarify the science of evolution but to promote a particular theology. Unfortunately, the challengers do this by misstating scientific facts, which I clear up in this part.

# Icons Used in This Book

The icons in this book help you find particular kinds of information that may be of use to you.

Sometimes, you can understand a scientific point by looking at it a little differently or comparing it to something similar, and this icon appears next to material that helps you do that.

This icon points out evolutionary principles that you want to remember because they're important to the topic at hand or because they're fundamental to understanding evolutionary biology.

This icon appears beside information that is interesting but not necessary to know. In fact, feel free to skip the info here, if you want. Doing so won't impair your understanding of evolution.

Sure, I could just tell you what evolutionary biologists say about evolution, but I prefer to show you how they know. This icon appears next to sections about scientific experiments designed to test evolutionary processes. And because the whole point of experiments is to test an idea, not to build up proof that the idea is right, some of these case studies end up revealing things that the researchers didn't necessarily expect.

# Where to Go from Here

This book is organized so that you can go wherever you want to find complete information. Want to know about the role natural selection plays in evolution? Head to Chapter 5. If you're interested in the link between individual fitness and how certain social systems develop, go to Chapter 11. You can use the table of contents to find broad categories of information or the index to look up more specific things.

If you're just beginning to learn about evolution, I suggest that you start with Part I. It gives you all the basic information about evolutionary principles and points to places where you can find more detailed information.

# Part I
# What Evolution Is

The 5th Wave                    By Rich Tennant

## In this part . . .

Evolution is the process that explains how organisms change over time. It's really as simple — and as profound — as that. And in this part, I begin with the very basics: key evolutionary principles, from Darwin's day to today. You'll also find a very brief discussion of genetics. Why? Because the only changes that are important to evolutionary study are genetic ones.

# Chapter 1

# What Evolution Is and Why You Need to Know

. . . . . . . . . . . . . . . . . . . . . . . . . . . . . . . . . . . . . . . . . .

## In This Chapter

▶ Understanding what evolution is

▶ Introducing the scientific field of evolutionary biology

▶ Realizing why evolution is relevant

. . . . . . . . . . . . . . . . . . . . . . . . . . . . . . . . . . . . . . . . . .

*E*volution. You've no doubt heard about it, and you've probably seen a show or two about it on TV, but its significance likely escapes you. Watching a bunch of scientists on the Discovery Channel dig in the dirt with little toothbrushes and get really excited about some little bit of bone or a tooth may leave you thinking, "Well, yes, those do look like teeth, and they certainly do seem old, but . . ." A tooth, you say to yourself, is hardly reason to trade high fives and uncork champagne bottles. At times like these, evolution can seem pretty slippery. After all, there's got to be more to it than a stray fossilized tooth or bone fragment.

Well, there is. Evolution explains how we (and I'm using *we* collectively to mean all living organisms: you, me, and all other animals; moss, trees, and the roses in your garden; viruses, amoebas, bacteria, and all the other little critters) came to be in all our complexity and variation. The reason scientists get excited about fossilized teeth is because findings like these are consistent with what scientists understand about the evolution of life on Earth. That single tooth is just one piece of the evolutionary puzzle; thousands more pieces exist. All together, those pieces form a picture of our genetic past and a road map that leads from a common ancestor to who, and what, we are today. It's a journey over billions of years.

This chapter gives you an overview of evolution in all its glory: what it is, how it works, and what it does. By the end, you may begin to understand what the great evolutionary biologist Theodosius Dobzhansky meant when he wrote, "Nothing in biology makes sense except in the light of evolution."

# Biological Evolution at a Glance

Evolution can be defined simply as change through time, and it can refer to anything that changes. Languages evolve; tastes evolve; cultures, art forms, and football offense strategy all evolve. This book isn't about evolution in general, though, but about biological evolution: the changes, over time, in organisms.

Biological evolution deals with a very specific type of change through time — changes in the frequencies of different genes — throughout an entire species, or within a single population of that species, from generation to generation. *Evolutionary biologists* — scientists who study evolution — just love that stuff. Their mission? To understand how evolution works (by figuring out what causes changes in gene frequencies) and what evolution does (by figuring out what sorts of things happen when gene frequencies change).

The following sections offer a general overview of how evolution works and what it does. Parts II and III delve into these topics in a great deal more detail.

## Gene defined

Back in Charles Darwin's day, a *gene* was defined simply as the unit of heredity. People knew that specific traits, such as blue eyes or red hair, were passed from parent to child, but they didn't know exactly what a gene was or how the process worked. Today, we know a lot more:

- ✔ We know about DNA *(deoxyribonucleic acid),* which is what gets passed from parent to offspring.

- ✔ We know that DNA is a long molecule made up of a string of four *subunits* (four letters); that the order of these letters, commonly called the *DNA sequence,* stores genetic information; and that a gene is a particular sequence of a particular piece of an organism's DNA.

- ✔ We've developed the chemical techniques that allow researchers to determine the exact sequence of an organism's DNA. As a result of this ability to work with DNA, scientists have a much better handle on the details of the evolutionary process.

What this means — and why it's important enough to include here — is that by being able to identify the DNA sequence of a particular gene, scientists can measure exactly what genetic changes occur across generations. Being able to measure things, especially things like DNA strands, gets evolutionary biologists all goose-pimply. (For more information about genes and DNA, head to Chapter 3.)

# What's the (gene) frequency, Kenneth?

Simply put, the *frequency* of a particular gene is how often it appears in a population. When researchers examine the DNA sequence at a particular location in a species' DNA in different individuals, they sometimes find that all the individuals have the same sequence. In this case, because only one gene (or one DNA sequence) exists at this location, its frequency is 100 percent. At other times, different sequences are present in different individuals. In this case, when more than one gene is present at this location, scientists speak of the frequencies of the different genes.

Suppose that you've discovered three different DNA sequences; call them genes A, B, and, C. If half the individuals you examine have gene A, one quarter have gene B, and one quarter have gene C, the frequencies of the three genes are 50 percent gene A, 25 percent gene B, and 25 percent gene C.

By identifying changes in the frequency of particular genes through the passing of generations, you can determine whether the organism has evolved. Using the example of genes A, B, and C from the preceding section, if you came back generations later to measure the frequency of these three genes again, and you found that the frequencies had changed, evolution has happened.

Here's an example: Suppose that you collect a bunch of a particular kind of bacteria and measure the frequency of the gene that makes the bacteria resistant to a new type of antibiotic. In your initial count, you find that the frequency of this gene is extremely low: Less than 1 percent of the bacteria have the gene that makes them antibiotic resistant. You come back in a few years. Your original bacteria are gone, but in their place are their great-great-great-great-etcetera grandkids, and you repeat the analysis. This time, you find that 30 percent of the bacteria have the antibiotic-resistant gene. Although you haven't actually witnessed evolution, you're looking at its result: the change in the frequency of particular genes over time. The antibiotic-resistant gene appeared in less than 1 percent of the original bacteria; it appears in 30 percent of the descendents. (Go to Chapter 17 for an in-depth discussion of the evolution of antibiotic resistance in bacteria.)

In a nutshell, biological evolution is simply a change in the frequency of one or more genes through time. Scientists collect this sort of data about the occurrence of evolution all the time — not only for bacteria, but also for all sorts of organisms, both simple and complex.

## The timescales of evolution

Although the changes in gene frequencies happen gradually through time, the rate of evolution isn't constant. Gene frequencies can remain constant for long periods of time and then change in response to changes in the environment. The rate of change can increase or decrease, but the basic process — gene frequencies changing over time — continues. To differentiate between these time scales of the evolutionary process, scientists use the terms *microevolution* and *macroevolution:*

- **Microevolution** refers to the results of the evolutionary process over short time scales and small changes. An example is a bacterium in a laboratory beaker experiencing a mutation that creates a gene that confers higher growth and division rates relative to the other bacteria and beaker. Microevolution, because it happens on a time scale that we're able to observe, tends to be a bit easier for us to wrap our brains around than macroevolution.

- **Macroevolution** refers to the results of the evolutionary process typically among species (or above the species level; see Chapter 11) over long periods. Nothing is different about the process; nothing special is happening. Macroevolution simply refers to the larger changes researchers can observe when evolution has been going on for a longer time and involves processes such as extinction, which may have little to do with microevolution. *Speciation,* the process whereby one species gives rise to two, is an example of macroevolution. Speciation isn't all that complicated, and scientists are getting a pretty good idea about how it works; you can find out more in Chapter 8.

Other than the time frame, no difference exists between micro- and macroevolution. The process isn't any different from what scientists can observe in a test tube in the laboratory (an example of microevolution); there's just been a lot more of it.

## Gene extremes: Mutation and extinction

Genes can go to extremes. At one extreme is the disappearing gene. Suppose that you measure the frequency of the three different genes at a particular site in a species' DNA and then return some years later to find that one of the genes is no longer present. That gene's frequency has dropped to zero. It's gone. It's extinct. When a gene goes extinct, the species that had the gene is still around, but at least at this particular location in its DNA, it's not as diverse.

At the other extreme, new genes can appear. The process by which the sequence of a parent's DNA is copied and passed on to the next generation is remarkably accurate. If it weren't, none of us would be here. But no process is perfect, and mistakes happen. These mistakes are called *mutations,* and they can result in a DNA sequence different from the original — in other words, a new, different gene. These new genes can affect the functioning of the organism in several ways:

- ✓ **They can have no effect at all.** Because there's a certain amount of redundancy in the code of the DNA sequence (go to Chapter 3 for the details), it's possible to change a letter here and there with no effect whatsoever. Even if the mutation does create a change, that change may not affect how the gene product functions. In both cases, the new genes don't have an impact — either positive or negative — on whether an organism survives.

- ✓ **They can result in a change that's harmful to the organism.** Most mutations that cause a change fall into this category. Even the simplest organisms are really quite complicated. If you change something randomly, most often the outcome is bad. Genes of this sort vanish as rapidly as they appear.

  Occasionally, bad mutations — which typically are destined for a short run before becoming extinct — actually *increase* in frequency. Here's how it could happen: If a gene with negative effects is present in the same critter as a gene with positive effects, the frequency of the bad gene can increase as it rides the evolutionary coattails of the really great new gene. Suppose that two mutations occur simultaneously in different locations on an organism's DNA: one resulting in a gene that is slightly harmful and another resulting in a gene that is advantageous. The slightly harmful gene may increase in frequency simply because it's along for the ride.

- ✓ **They can result in a change that's advantageous to the organism.** This class of mutations is by far the rarest, but beneficial mutations do occur. These mutations, although rare, can increase in frequency. Ultimately, they're the source of all the variation upon which evolution by natural selection acts. (Skip to Chapter 4 for more detail about the role variation plays in evolution.)

All the different genes in all the organisms on earth started out as mutations that, though initially rare, ended up increasing in frequency. As the source of new genes, mutations are a key part of the evolutionary process. A gene can't increase or decrease in frequency until it first appears, and mutations are how that happens.

# Darwin and His Big Ideas

You can't talk about evolution without talking about Charles Darwin (1809–1882), a would-be physician and theologian whose fascination with natural history and geography led him to accept a position as gentleman's companion to the captain of the *HMS Beagle,* a ship bound for South America with the purpose of mapping the area and sending plant, animal, and fossil specimens back to England. The voyage lasted five years, from 1831 to 1836.

Several things led Darwin to speculate about the changes that might occur in species over time: the diversity of life he observed on his voyage, the geographical patterns whereby different yet obviously related species were found in close proximity to one another, and the fossils he collected that made it clear present-day species weren't the ones that had been present in the past.

Darwin returned to England in 1836, already well known in the scientific community for the specimens and detailed notes that he had sent back. By 1838, Darwin had developed in more detail his theory of how gradual changes resulting from natural selection could result in changes in existing species as well as the formation of new ones. Over roughly the next 20 years, Darwin continued to develop and refine his ideas. In 1859, he published his seminal work, *On the Origin of Species,* which laid out the foundations of evolutionary theory. The following sections hit the highlights of Darwin's ideas. His other works include *The Descent of Man, and Selection in Relation to Sex* (1871) and *The Expression of the Emotions in Man and Animals* (1872).

Find the title *On the Origin of Species* a bit cryptic? Roll the full title around your mouth for a while: *On the Origin of Species by Means of Natural Selection, or the Preservation of Favoured Races in the Struggle for Life.* The shorter version may not be as descriptive, but it certainly is easier to remember — and say! You can read it (and all of Darwin's other works) at http://darwin-online.org.uk/.

Darwin didn't use the word *gene;* instead, in his work, he referred to *characters.* Yet because his ideas focused on *heritable* characters (that is, those that can be passed from parent to offspring), his "characters" are directly linked to genes.

## *Natural selection*

One of Darwin's big ideas was what he called *natural selection,* the mechanism that he proposed to explain what he called "descent with modification" — that is, changes in an organism through subsequent generations. (Today, we'd say that natural selection explains how gene frequencies could change over generations.) This big idea, explained in depth in Chapter 5, is both remarkably insightful and remarkably simple, which explains why it's stood the test of time.

Basically, Darwin recognized that some characters get passed from one generation to the next and others don't. What he wanted to understand was *how* descent with modification could have occurred. What was the underlying driving force? He concluded that the driving force was the process of natural selection: Not all individuals in a given generation have an equal chance of contributing to the next generation. Some are selectively favored; some are selected against.

Darwin surmised that natural selection worked the same way as the process of artificial selection used in animal husbandry and agriculture:

- **Artificial selection:** Since before Darwin's time, people have been selectively breeding animals and plants: chickens that lay more eggs, cows that make more milk, pansies that are brighter and last longer . . . and the list goes on. Essentially, humans have been pretty apt hands at spurring evolution in agriculturally important plants and animals. *We* decide which genes are more likely to make into the next generation. The cows that produce more milk are the ones that we selectively breed to produce better dairy cows; the ones that make less milk, we eat. As a result of the choices we humans make as selecting agents, we can dramatically alter in a relatively short period the characteristics of the organisms we breed.

- **Natural selection:** Darwin realized that if humans, by the process of artificial selection, could create such major differences over the extremely short period of time, then the natural environment, acting over a much longer time scale, could have produced much larger changes. Darwin called his process *natural selection* because the natural environment, not humans, was the selecting agent.

In artificial selection, farmers and breeders determine which characters they like and work to propagate in their produce and livestock. In natural selection, the same type of selection occurs, but the selecting factor isn't man, but nature, or the environment in which the organism exists. To help you understand the difference between artificial and natural selection, consider the cow. In the barnyard, farmers selectively favor the cow that makes the most milk; in the wild, natural selection favors the cow that can make enough milk to feed its calf and still do all the other things the cow needs to do to survive on its own.

Whether natural selection favors an individual is a function of the individual's particular heritable characters. Some heritable characters increase the probability that the individuals containing them will contribute to the next generation; some characters decrease the probability that individuals will contribute to the next generation. What all this means is that organisms in the first category reproduce more than do the organisms in the second category. *That's* what makes one generation different from the next, the next different from the one that follows it, and the one that follows it different from the one that comes later . . . and so on and so forth, ad infinitum.

# In case you're curious: Survival of the fittest

Although he didn't coin the phrase *survival of the fittest,* Darwin did make it a household term. Many people assume (erroneously) that it means the natural order mandates that the strong survive and the weak die away. But to Darwin and other evolutionists, *survival of the fittest* is simply synonymous with *natural selection.* In other words, those organisms that possess selectively favored heritable characters are the ones that pass their genes into the future with the most success.

This sidebar marks the first and the last time you'll see this phrase in this book. Why?

✔ **It's problematic.** The phrase doesn't clarify the concept Darwin was trying to explain (although he no doubt thought it did; otherwise, he wouldn't have used it). To express the concept more clearly, Darwin could

have used the term *survival and differential reproduction of the fittest,* but that's just not as sexy.

✔ **It doesn't make much sense semantically.** In beginning a study of evolution, students often say, "Well, if evolution is survival of the fittest, and the fittest are the ones that survive, that seems pretty circular." Indeed, it is.

✔ **Even evolutionary biologists never use it.** They use *natural selection* instead.

Don't let any of this get in the way of your developing an understanding of the term *fitness,* however; that word is crucially important to understanding evolution. Head to the section "How 'fitness' fits in with natural selection" for details.

Here's an example: Imagine a population of lions. Half the lions have the work-hard-run-fast-and-catch-lots-of-gazelles character. The other half have the sit-around-and-be-lazy character. It's tough in the Serengeti, and only the lions with the work-hard-run-fast character manage to store up enough energy to reproduce and raise offspring successfully. If you reanalyze this population after a few generations have passed and find fewer lions with the lazy character, that's evolution driven by natural selection!

# Speciation

Darwin realized that because individuals differ in the characters they have, and because these differences affect their chances of survival and reproduction, some characters are more likely to get into the next generation than others. He also realized that as a result of this process, the frequency of characters changes over generations. Pass through enough generations, and the sum of all the little evolutionary changes may result in an organism that's evolved into an entirely different species.

Here's a quick example: Imagine you have two populations of the same animal. Each population lives in a different place, and the populations rarely interbreed. The selective forces in those two places — the combination of things we call the environment — is different. In one environment, it's good to have a long beak; in the other environment, a short beak is better. Other significant environmental differences exist as well. It's very wet on one side of the mountain range and very dry on the other, for example. In two such different environments, gene frequencies change in one way in the first location and another way in the second. Over a long period, the two populations become so different that they can no longer interbreed. They have become different species.

Today, scientists can identify all the stages of speciation in the natural world. They can find pairs of species that seem to have diverged from a single species very recently, and they can find pairs of populations that appear to be on the verge of becoming separate species. In some cases, the two populations are so close to becoming different species that all it would take is some minor habitat change to push them that last little bit and turn one species into two. For more detailed information about speciation, head to Chapter 8.

The idea of speciation got Darwin into a lot of hot water, and it's a hot-button issue today because it links organisms to common ancestors, which is all well and good for things like fish, oak trees, and invertebrates. But when you throw humans into the mix — whoa, Nelly. To read more about the conflict between evolutionary science and those who deny it, head to Chapter 22.

## *How "fitness" fits in with natural selection*

The process of evolution by natural selection is driven by differences in *fitness,* or how successful an organism is at getting its genes (or characters) into the next generation. In short, fitness is all about how well an organism reproduces. Characters (or genes) that increase an individual's fitness are more likely to be passed to the next generation than genes that don't. This process is how the frequency of genes changes through time.

In the evolutionary process, fitness has nothing to do with how buff you are. It's purely a measure of the differential reproductive success among different individuals, which is a fancy way of saying that it refers to how successful an individual is at producing offspring. If one individual produces twice as many offspring as the next individual, all other things being equal, it's twice as fit.

## *Understanding adaptive characters*

Some evolutionary changes are *adaptive,* meaning that a character has changed as a result of natural selection in a way that makes that character better suited to perform its function. Here's an example of the process of adaptation: Gazelles run away from cheetahs. The slow gazelles get eaten, leaving the faster gazelles to reproduce. In the next generation, the gazelles are faster on average than those in the past generation, because the run-away-from-cheetahs character has evolved. Being able to run really, really fast is an adaptation.

It's not always easy to tell whether a particular character is an adaptation because sometimes things that appear to be adaptive characters aren't. Suppose that you have a cat and decide to put its food outside. At some point, you notice that birds eat the cat food. Knowing a bit about evolution, you think that eating from the cat dish may well be good for the birds; they probably have more energy to sing songs, build nests, and raise baby birds. If you observe such successful foraging behavior in a different environment, you might conclude that the birds are foraging in your cat dish as the result of natural selection. But eating cat food isn't an adaptation (the birds haven't evolved to eat out of cat dishes); it's opportunistic. The food's available, and the cat . . . well, he's probably trapped behind a patio door. For more about adaptive characters, go to Chapter 5.

# The Study of Evolution, Post-Darwin

Darwin had only a vague idea of what genes were and didn't know squat about DNA, but he hit the evolutionary nail on the head. Today, scientists know that the process of evolution by natural selection occurs pretty much the way Darwin first proposed it: Natural selection results in changes over time in any given population, and good genes (those that make the organism more fit — that is, more successful at surviving long enough to reproduce) become more frequent over time. Still, scientists' understanding of evolution has continued to evolve as they expand the theory of evolution to include some elements Darwin was unable to address:

- ✔ **Many DNA mutations are selectively neutral.** The DNA code contains a certain amount of redundancy, which means that many changes in the DNA don't result in a fitness advantage or a fitness cost. The extent to which these genes increase or decrease in a population has entirely to do with chance.

- ✔ **Chance can be an important factor contributing to the change in gene frequencies through time.** Imagine that half the deer in the forest have blue eyes, and half have brown eyes. Now suppose that a couple of trees fall over and accidentally crush a couple of deer with blue eyes. All other things being equal, the next generation will have a higher proportion of the brown-eyed gene than the previous generation. Evolution has happened, but *not* as a result of natural selection. (Yes, I know that deer don't have blue eyes; it's just an example.) For more information on how chance factors into the evolutionary process, head to Chapter 6.

I can imagine what you must be thinking: Two deer more or less are hardly going to make much of a difference. In a large population, you'd be right, but in a small population, a few deer more or less can make a difference that would be noticed in the future. When the population is large, chance events aren't as important, but when the population is small, random events can have larger repercussions.

✔ **Not all the characteristics of any particular organism are positively correlated with fitness.** This idea stems from scientists' understanding that not all evolutionary change is the result of natural selection. Sometimes, it's the result of chance; sometimes, it's the result of bad genes hitching a ride into the future with the good genes that made the organism more fit.

✔ **The environment affects fitness.** Populations in different places experience different selective forces. A gene for being able to survive a long time without water, for example, may offer a fitness advantage in the desert, but it may have rather negative consequences in a rain forest. Interaction between the gene and its environment is important in determining whether a given gene increases or decreases fitness.

Sickle cell anemia is an example of how the environment determines whether a particular gene increases or decreases fitness. The gene that causes sickle cell anemia produces a slightly different form of hemoglobin. The most extreme case occurs when someone has two copies of the sickle cell gene: one from the mother and one from the father. But even having just *one* sickle cell gene causes illness. At first glance, it seems obvious that this gene wouldn't increase anyone's fitness, yet it's present in high frequencies in certain areas of Africa. By examining the system from an evolutionary perspective, scientists learned an interesting thing about the sickle cell gene: Having a copy of this gene helps protect against malaria, which is present in those areas of Africa where the gene occurs at high frequency. So yes, it's bad to have this gene in the current era in the United States. But in the days before antimalaria drugs, it was a good gene to have in parts of Africa.

# Applying Evolution Today

Evolution is interesting purely for its own sake, but of course, I *would* think that, having devoted years to studying, teaching, and writing about it. But evolution is good for more than just student lectures and small talk in academic circles: Understanding what evolution is and how it works makes all sorts of things possible. The following sections give a small sampling of how scientists apply aspects of evolutionary biology. You can find many more examples throughout this book.

## Conservation

Understanding evolution helps conservationists in their efforts to protect endangered species. When resources are limited, as they often are, scientists have to make choices about which natural areas to protect and which populations of species to focus on. Understanding evolution can help them decide where to devote resources.

For example, many people think that the key to protecting endangered species is to conserve the maximum number of individuals possible. But understanding evolutionary biology and the patterns of variation present in natural populations helps us recognize that the real key is conserving genetic variability. If two populations are genetically different, part of a viable conversation management plan is maintaining this diversity, for two reasons:

- ✔ This diversity is a characteristic of the species that the scientists are trying to protect.

- ✔ The naturally existing variation allows the species to respond to future changes in the environment.

Another thing that evolution teaches — specifically, evolution by random events — is that we can't allow endangered populations to reach critically low numbers. In small populations, the variations scientists are trying to conserve — the very essence of what makes a particular species unique — are at risk of being lost due to random events that would be insignificant in a larger population. (For more information on the role chance plays in small populations, go to Chapter 6.)

## Agriculture

Although humans have been breeding plants and animals for thousands of years, recent understanding of the evolutionary process lets us attack this task in a more scientific fashion. Following are some highlights in the field of agriculture, courtesy of our understanding of the evolutionary process and principles:

- ✔ **Advancements in breeding:** Understanding the detail of the evolutionary process can help us devise new breeding strategies. Head to Chapter 11, which explores in detail a breeding program that successfully bred chickens that produced more eggs by selecting for chickens that got along well together in chicken coops — definitely not the normal situation and something that had been a serious problem in chicken farming before these developments.

- ✔ **Crop variation:** The presence of genetic variation allows populations to respond to environmental changes; in the absence of such variation, populations can be destroyed by a sudden environmental change. Plant genetically similar crops over wide areas, and you run the risk of an agricultural disaster. Case in point? The Irish potato famine. Across Ireland, genetically identical potato plants were cultivated; a disease that attacked one potato turned out to be able to destroy them all, with horrific results.

✓ **Crop history:** Evolutionary biology allows scientists to understand the history of crop plants. Corn, for example, was domesticated by Native Americans, but for the longest time, biologists had no idea what wild plant it was derived from. Now, detailed studies of the evolutionary relationships of plants allow scientists to identify the wild plant from which corn was artificially selected. Having found the parent plant, scientists can study the genetics of how this plant survives in the presence of insects and microbial pests, which can only help in the quest to develop even better corn.

# Medicine

The field of evolutionary biology affects the medical profession in three key ways: figuring out what has happened, understanding what is happening now, and trying to predict what will happen in the future to human disease. All three help researchers devise strategies for prevention and treatment of health problems big and small.

One area of particular medical importance is the evolution of microbes — the viruses, bacteria, and other microscopic critters that cause infection — that are increasingly resistant to antibiotics. The more researchers know about how and why microbes evolve as they do, the better they'll be able to counteract the effect of those microbes. Consider, for example, the virus that causes AIDS. Reconstructing evolutionary history has allowed researchers to trace the spread of human immunodeficiency viruses (HIV) across the globe, as well as to determine the relationships among human viruses and the immunodeficiency viruses of other animals. From these studies, scientists know that these viruses don't always cause disease in their hosts. By studying related harmless viruses, researchers may be better able to understand exactly why HIV is so dangerous in humans.

The study of evolutionary biology also guides treatment of diseases. The highly successful triple drug therapy that's been amazingly beneficial to HIV-positive individuals is the direct result of scientists' knowledge of how antibiotic resistance works in microbes: Even though mutations in the HIV virus render it resistant to medications, it's more difficult for the virus to evolve resistance to all the drugs at the same time. Finally, by examining how HIV evolves resistance to medicines, scientists hope not only to design better medicines, but also to identify how best to design a vaccine.

Chapters 17, 18, and 19 are chock full of information about the role of evolutionary biology in the fight against disease.

# One Final Point: Just How Evolved Are You?

Evolution isn't a race to some cosmic finish line. No species is more evolved than the next. Every living thing is descended from the same common ancestor. All the different lineages have been evolving for exactly the same length of time. True, humans are better than pine trees at doing the things humans do, but we can't stand outside in the sun and soak up energy — something that pine trees do very well. The reason life is so different is that different environments select for different outcomes.

Neither is evolution a climb to the top of some life-form ladder on which the "higher" orders take over the top rungs (we humans are at the tippy-tippy top) and the "lowlier" creatures hang around the base. In fact, not all evolution results in more complex life forms. This point may seem like a small one, but it's actually quite important and is easily lost when most people think of evolution in terms of the "monkey-to-man" graphic — the one that shows the evolution of man in a series of stages, from monkey to ape to caveman to investment banker. Although you can make an argument that the caveman gave way to the investment banker and therefore forms a valid time series, other primates are still around and are just as evolved as humans are.

Evolution *can* lead to greater complexity, but it doesn't always. Over the history of the earth, since the first single-celled life forms, there was really nowhere to go but up in terms of size and obvious physical complexity. But as soon as larger, more complex critters evolved, the possibility existed that some would evolve simpler forms. Parasites, for example, have lost many of the functions that they can scam off their hosts. The eyes of cave-dwelling organisms constitute another example. Absent the need for the complex structure of the eye, mutations that cause a reduction in the eye can pile up.

P.S: Just between you and me, I do sometimes think of myself as being a bit more evolved than a bacterium — but then I think of the incredible biochemical diversity that bacteria are capable of, and I realize the error of my ways.

# Chapter 2

# The Science — Past and Present — of Evolution

. . . . . . . . . . . . . . . . . . . . . . . . . . . . . . . . . . . . . . . . . . . . . . .

*In This Chapter*

▶ Getting familiar with the language of science

▶ Digging into fossils and rocks

▶ Comparing Darwin's knowledge with ours

. . . . . . . . . . . . . . . . . . . . . . . . . . . . . . . . . . . . . . . . . . . . . . .

*T*oday, scientists know a lot more about fossils than they did back in Charles Darwin's day. They understand the molecular mechanisms of genetics and heredity. They can conduct experiments in the laboratory that allow them actually to observe the evolutionary process. In essence, science has come a long way since Darwin, whose investigations focused primarily on the process of natural selection. As this book explains, there's more to evolution than just the process of natural selection, but evolution by natural selection is an extremely important evolutionary force.

When it comes to natural selection, Darwin got the basics right, even though he had to speculate about the things that researchers can just measure today. Scientists since Darwin, in test after test and experiment after experiment, haven't been able to refute Darwin's theory. In fact, their work has provided copious evidence that the evolutionary process works almost exactly the way Darwin speculated it did.

This chapter looks at the information Darwin had when he formulated his theory of evolution via natural selection and at the things scientists have learned since then. The chapter also helps you understand exactly what scientists mean when they say *fact* and when they say *theory*, which, believe it or not, isn't code for "pulled out of thin air."

# Evolution: A Fact and a Theory

One of the most common sources of confusion for people trying to understand evolution is the question of whether it's a fact or a theory. Here's the answer: It's both. The *fact* of evolution refers to the things scientists can see and measure. The *theory* of evolution refers to the intellectual framework science has developed to explain the underlying processes that account for those facts.

To help you understand how evolution can be both a fact and a theory, I talk about another natural process that is both fact and theory: gravity. And because a key to understanding anything you hear or read about evolution requires being familiar with scientific principles, I also offer a short and sweet explanation of how scientists think.

## Evolution and gravity: Two peas in a scientific pod

The fact of gravity is beyond dispute. When you drop something, it falls — a fact that most (dare I say all?) of us have personally established, either intentionally or accidentally. We not only know that things fall, but we also know a few details about the falling process. We know that whatever we drop falls toward the Earth, and we can measure the downward acceleration. We know that the pull of gravity is different away from Earth. On a smaller body such as the moon, things don't fall as fast. In deep space, far from the Earth and the moon, things don't fall at all; they just float. The strength of this *attractive force* (a fancy way of saying *gravity*) has to do with the mass of the attractive body and the other object's distance from it. The moon, for example, which is smaller than Earth, has less gravity.

Humans know all these things about gravity, but it turns out that we're not exactly sure what gravity is or how it works. Why, for example, does a dropped object go down rather than up? Scientists in the field of physics called gravitational theory are trying to figure these things out, and they continually fine-tune their theories as more information becomes available. But just because they can't say definitively what gravity is doesn't mean that gravity is any less real.

Just because you and I don't understand gravitational theory doesn't affect how we interact with gravity on a daily basis. We know it happens even if we don't know why, and we take it into consideration when we launch things into space, skateboard, land an airplane, or try to get that nine-iron shot onto the green. The same is true of evolution: We have the facts of evolution (the aspects of heritable changes in living organisms that we can see and measure), and we have the processes that evolutionary biologists theorize are responsible for these facts. As it turns out, compared with their theories about gravity, scientists have an excellent understanding of the basic process of evolution.

# How to think like a scientist

Science is by definition a very conservative discipline, and scientists are extremely hesitant to say that they are certain about *anything*. In fact, one of the rules of science is that you must always allow for the possibility that additional data will force you to let go of an idea that you were really sure about and quite possibly very fond of. Scientists take this rule *very* seriously — so seriously, in fact, that before they get their diplomas and lab coats, they have to pinky-swear that they'll follow it.

Suppose that you want to know whether a particular coin is a fair coin, that is, as likely to come up heads as tails when you flip it. So you flip the coin a few times, and it always comes up heads. The fact that you got heads in your first few flips may lead you to suspect that the coin isn't fair, but because you flipped it only a few times, you can't say for sure. What you have is a suspicion. What you need is more information. So you proceed to flip the coin all day long, and the next couple thousand times you toss it, the coin always comes up heads. With a few thousand flips under your belt and a head coming up each time, you conclude that the coin isn't fair. This process seems straightforward, and the conclusion seems obvious — to anyone who isn't a scientist.

But a scientist in the exact same position and with the exact same data would not say that the coin isn't fair. She would cite a very, very, very low probability that it's a fair coin, or she'd say that she's 99.99999 percent sure that the coin isn't fair. The reason for the difference is the precision with which scientists state what they know. After all, it *is* possible that a fair coin could come up heads that many times in a row; that possibility just isn't very big.

To which the layperson may reasonably respond, "So you're not really sure!" What the scientist is really saying is that she *is* sure — within a reasonable doubt. (If that explanation's not good enough for you, consider the shaky ground on which it puts the entire American justice system, which also relies on the standard of reasonable doubt.)

### You say toMAYto; I say toMAHto: The language of science

Do scientists ever seem as though they speak their own language? In a way, they do. You probably recognize some words as scientific terms — words like *nucleotide, paedomorphosis, heterozygosity,* and others, which seem to be little more than Latin and Greek roots randomly strung together. But other terms — ones that mean one thing to laypeople and something else to scientists — are a bit trickier. The subtle differences in the way scientists and nonscientists use the same words are often sources of confusion.

You won't find a better example of this situation than the word *theory*. In scientific terms, a *theory* is a hypothesis that has overwhelming support — in essence, an idea that's been proved. In layman's terms, *theory* typically means *best guess*. See the problem?

To help you understand the scientific meanings of science's three most important words (the fourth is *funding*), I offer these definitions:

- ✔ **Fact:** Something you can observe or measure.

- ✔ **Hypothesis:** A working idea or set of ideas resulting from observations and measurements. The hypothesis serves to guide future investigations. It gives scientists suggestions about what facts they should try to collect next.

- ✔ **Theory:** A conceptual framework, tested repeatedly but not rejected, that explains the facts, observations, and measurements, and makes accurate predictions of how the system will behave in the future.

The facts of evolution, as I show throughout this book, are clearly established. The current theories of how evolution functions are solidly supported as well. But the linguistic difficulties in communication are such that people continue to ask scientists, "But it's just a theory, right?"

### Scientific investigation

Scientists don't start talking about something as a theory until it has overwhelming support. How do they get that support? Through scientific investigation. A coin toss is a good example of the process of scientific investigation:

1. **Start with some observations about the natural world.**

   In the example of the coin toss, you observe that the first few tosses always end up heads.

2. **Formulate a hypothesis.**

   Your hypothesis after flipping head after head? That the coin isn't fair.

   The hypothesis serves as the scientist's starting point; maybe it's right, and maybe it's wrong. They key is to do enough testing to find out.

3. **Gather additional data to test this hypothesis.**

   In the coin example, you gather additional data by tossing the coin several thousand more times.

   Repeating one type of test ad infinitum — exhaustively flipping a coin, for example — and getting a particular result isn't good enough. First, you have to address and eliminate any other factors that could affect the test results. Maybe the coin is fine, but something about the person flipping it isn't quite right. Or maybe the coin is fine, but some other factor — like an air vent blowing air over the researcher's head — keeps it from landing heads or tails half of the time. You need to have the coin flipped by a bunch of other researchers in other parts of the room to make sure they get the same result.

4. **As your data accumulates, it either supports your hypothesis, or it forces you to revise or abandon the hypothesis.**

   In the coin-toss example, heads continue to come up, thus lending support to the original hypothesis that the coin isn't fair.

5. **At the point where an overwhelming amount of information starts to accumulate in support of the hypothesis, the hypothesis is elevated to a theory.**

The hypothesis must get tested and tested to the best of everyone's ability before it arrives in the exalted land of theory. And even a theory is only one good experiment away from being rejected, which is one of the fundamental components of the scientific method: that the ideas scientists come up with must be *falsifiable*. That is, scientists must be able to imagine some set of results that would cause them to reject the theory; then they must see that over and over again, they never get the expected results. This process always sounds somewhat backward to nonscientists, but that's just the way scientists do things. We scientists never say that an idea is true; what we say is that, even after our best efforts, we've been unable to show that it is false. Then it's high fives all around, and we go grab a beer.

# The Evidence of Evolution

Evolution is simply the change over time in the frequencies of different heritable traits (that is, the frequencies of different genes) in a group of organisms. One year, half the birds have blue eyes and the other half have brown. Some years later, two thirds of the birds have blue eyes. If eye color is a heritable trait, that is, one that is passed from parent to offspring, then that change is evolution.

Darwin's big idea was coming up with a process that could cause these changes. This theory of natural selection is simply a process in which heritable traits that help an organism survive in its current environment occur more frequently in subsequent generations. In other words, *genes* (the heritable traits) that are *favored* (help survival) increase over time. Evolution by natural selection really is that simple, regardless of what you may have heard to the contrary.

Darwin, the father of evolutionary theory (refer to Chapter 1), did a phenomenally good job of piecing this part of evolutionary process together, but he could only hypothesize about the underlying details of what he called *descent with modification*. What he didn't understand was the underlying genetic mechanism that was responsible.

## Knowledge about DNA and genetics

Making out heritable traits just requires knowing a little bit about DNA, genes, and the genetic code — topics covered in more detail in Chapter 3. This knowledge of genetics allows researchers to measure gene frequencies. Today, scientists can do more than just observe that evolution has happened; instead, they can determine the specific genes involved and measure the rates at which the genes' frequencies change in populations where different selective forces are at play. Scientists' knowledge of exactly what is going on gives them a much richer understanding of the process and allows them to ask far more detailed questions than Darwin could.

## Experimental evidence

Darwin saw evidence that the process of natural selection had occurred in nature, but he was unable to watch it happen. Scientists today, however, can actually observe the evolutionary process in action. Using organisms that reproduce rapidly and can be kept in laboratories in large numbers (such as fruit flies and bacteria), and employing modern tools and techniques (such as DNA sequencing), evolutionary biologists can conduct actual evolution experiments and watch the results.

Scientists don't pick fruit flies because anything about that species is special; they pick fruit flies for expediency. Experiments conducted with fruit flies go a lot faster than experiments with other organisms (such as dogs or sea turtles) because the generation time of the flies is much, much shorter. A fruit fly is born, reproduces, and dies within a couple of weeks. This short life span gives scientists more opportunity to examine changes in heritable traits.

## Measurement of the rates of change

Scientists can measure the rates at which evolutionary changes happen. By measuring changes in the frequencies of existing genes and observing mutations, they can describe the evolutionary process exactly. They can see in the laboratory how selection can result in changes in gene frequencies.

Here's a favorite example of mine to show how witnessing evolution in action is possible: antibiotic resistance. Scientists can observe how the frequencies of genes that make a microbe resistant to antibiotics increase when the environment changes to include an antibiotic. They take a beaker full of microbes and add some antibiotics. Then — big drumroll here — they come back the next day and find that the frequency of genes conferring resistance to antibiotics is much higher than it was before the antibiotics were added. Why? Because all the bacteria that didn't have antibiotic-resistance genes died and the ones that did have antibiotic resistant genes reproduced a lot! Hence, the favored trait — resistance to antibiotics — increased over time. You can read more about antibiotic resistance in Chapter 17.

Mutations don't occur in *response* to environmental change. Instead, the mutations already exist and are favored in the new environment. Beakers full of bacteria tend to contain mutants that are resistant to antibiotics even when no antibiotics are around. Toss an antibiotic into the mix, and the mutation gets its chance to shine by enabling the bacteria to survive in the presence of the antibiotic.

# The Scientific Foundation of Evolution by Natural Selection

Charles Darwin observed that the offspring of a particular parent, although they resembled the parent, tended to differ from the parent in various ways. That is, the offspring were *variable*. Based on his observations, Darwin hypothesized that, because of their inherent differences, some of the offspring would be better than others at doing whatever it is they needed to do to survive and reproduce. Further, he surmised that if the differences that resulted in increased survival and reproduction were *heritable* (that is, passed from parent to offspring), they would be passed disproportionately to the next generation, and through time, this process would lead to changes in the species.

Darwin didn't pull his ideas out of thin air. He developed his theory of evolution during the period when rapid advances were being made in a variety of fields, including geology, selective breeding in agriculture, and *biogeography* (the study of the locations of different species).

Not surprising, scientists have learned a fair bit more about the natural world in the hundred-plus years since Darwin proposed his theory of evolution by natural selection. What *is* surprising is how well most of what researchers have learned since Darwin has been in agreement with his hypothesis. As scientists have developed a more complex understanding of the details of the evolutionary process, their confidence has only increased that the mechanism Darwin first proposed is correct. This section outlines what Darwin knew and some of the things scientists have learned about the evolutionary process since Darwin.

## Gradualism: Changes over time

Although people now take for granted the idea that gradual processes acting over long periods can have dramatic effects — think, for example, of the slow

erosion by the Colorado River that led to the formation of the Grand Canyon —
this idea was at odds with the prevailing view in the 1800s that the Earth was
very young. Then along came the field of *stratigraphy,* which deals with the
horizontal banding patterns that you can observe in the faces of cliffs or when
a highway is cut through deep rock.

By Darwin's day, detailed geological mapping of Europe had revealed that a
reproducible sequence of bands was spread across a large geographical area
and that these bands contained fossils. Even in the absence of detailed information about the absolute ages of the different bands, scientists concluded
that the ones on the bottom were typically older than the ones on the top.
The very existence of these bands and the fossils that were found within
them hinted at a process of gradual change.

If the new geological views about gradualism were correct — that is, that
the Earth formed over long periods, as indicated by the banding patterns of
different geological eras — scientists could imagine that the changes in the
biological community were also the result of small changes occurring over
a large period. Turns out that they were right on both counts.

## The age of the Earth

Although earlier scientists didn't have the tools to date the age of the Earth as
we do today, they understood that the lower bands (and the fossils in them)
were older than the higher bands and their fossils. Still, Darwin had no idea
how immensely old the Earth was or how long the evolutionary process had
been going on. Even when people began to understand that the world was
quite a bit older than previously thought, they couldn't give an exact age to it.

## Dating the age of the Earth: Radioisotope dating

Scientists know that the Earth is about 4.5 billion years old by using radioisotope dating techniques. To understand how this process works,
you need to know a little bit about atoms and
isotopes. For those who need little refresher
course on basic chemistry, think of water, or
$H_2O$. The *H* and the *O* refer to hydrogen and
oxygen, the two atoms that make up water. As

the notation indicates, water consists of one
molecule of oxygen and two molecules of
hydrogen.

Often, any one atom has several different forms,
which are referred to as *isotopes.* Atoms are
made up of electrons, protons, and neutrons,
and the number of electrons and protons

determines the type of atom. Hydrogen, for example, has one electron and one proton. Sometimes, it also has a neutron. The term *heavy water* refers to water in which each hydrogen atom has a neutron. This isotope of hydrogen is also called *deuterium.*

Some isotopes, like deuterium, are *stable,* which means that they're perfectly happy with the number of electrons, protons, and neutrons they have. Other isotopes are *unstable* because the different number of neutrons interacts with the other atomic components in such a way that some of the bits go flying off and, over a period of time, the isotope changes into some other atom. When these unstable isotopes change to a different atom, they emit radioactivity. For that reason, they're called *radioisotopes.*

An important property of radioactive isotopes is that scientists can describe very accurately the average probability of the transition's happening and express that probability as a number called the *half-life* — the time it takes for half of the atoms to undergo this transition. In the first half-life, half of the atoms transition. In the second half-life, half of the remaining atoms transition, leaving one quarter of the original parent material. In the third half-life, half again transition, leaving one eighth, and so on. (*Remember:* Just because half the isotopes decay in the first half-life doesn't mean that the other half decay in the second half-life—you'd be surprised at the number of students who make this assumption. Only half decay every half-life.)

To determine the age of material, researchers compare the ratio of the parent and daughter products that were initially in the sample with the ratio of these products at the current time. By doing so, they can calculate how much time has passed. The atomic clock is a very accurate national timekeeping apparatus calibrated by the precise regularity of radioactive decay.

Numerous radioactive isotopes exist. One system that has been very successful in dating the ages of fossils is potassium-argon dating. Potassium is an extremely common element. Although most potassium isotopes aren't radioactive, one of them is, and one of its decay products is the gas argon.

Potassium–argon dating relies on the fact that although potassium is a solid, argon is a gas. When rock is melted (think lava), all the argon in the rock escapes, and when the rock solidifies again, only potassium is left. The melting of the rock and releasing of any argon present set the potassium–argon clock at zero. As time passes, argon accumulates in the rock as a result of radioactive potassium decay. When scientists analyze these rocks and compute the ratio of argon to potassium, they're able to determine how long it's been since the lava cooled. When scientists date rocks from our solar system this way, the oldest dates they find are 4.5 billion years.

No fossils are present in lava, obviously; anything that was there melted along with the rock. But by dating the lava flows above and below a fossil find, scientists can put exact boundaries on the maximum and minimum age of that fossil. In this case, the variation in possible ages of the fossil simply reflects the fact that the fossil exists between the dated lava flows.

Radioactive dating has been perfected to the extent that scientists can get within a few percentage points of the actual date. They know this because they're able to date lava flows that happened recently enough for their dates to be known historically. Potassium–argon dating has been used to date accurately the age of the eruption of Mount Vesuvius at Pompeii, for example. The scientists knew that the technique worked because the age their equipment indicated matched the age noted in historical Roman records.

## Types of rocks and fossils

Geologists in Darwin's day were familiar with the diversity of types of rocks, but they were only beginning to appreciate the vast time scales over which geological events occurred. They had yet to understand the dynamic nature of the Earth's crust, and they lacked modern understanding of how these types of rocks formed:

✔ **Igneous:** Igneous rocks are of volcanic origin. They form when molten lava from volcanoes cools and solidifies. Basalt and obsidian are examples of igneous rocks.

✔ **Sedimentary:** Sedimentary rocks are formed by the gradual deposits of sediments. Sandstone is an example of sedimentary rock.

✔ **Metamorphic:** Metamorphic rocks are rocks of any origin that have been subjected to the extreme stresses and temperatures caused by the folding and crushing of the Earth's crust.

Understanding these rock types helps biologists understand the fossil record. Fossils are found only in sedimentary rocks. The molten lava that bubbles up from beneath the Earth's crust during a volcanic eruption doesn't contain any fossils (whatever had been there would have melted in the molten rock). Metamorphic rocks — even those of sedimentary-rock origin — don't contain fossils, because the extreme temperatures and pressures that converted the rock from sedimentary to metamorphic would have destroyed whatever fossils may have been there.

Today, scientists know quite a bit more. First, through radioactive dating, a painfully complex process whose details you don't need to worry about, they know that the Earth is about 4.5 billion years old. Scientists also know that life has existed on Earth for at least 3.5 billion years — a number that keeps changing as older and older fossils are found.

Although the age of the Earth may seem to be somewhat unrelated to evolution (rock, stone, and tectonic plates aren't living organisms and, therefore, don't "evolve"), it's actually very important to the theory of evolution because biological evolution needs time to happen. By knowing the actual age of the Earth and how long life has been present, scientists can ask whether enough time has passed for simple creatures such as the ones they see in the oldest rocks to evolve into more complex creatures, such as the ones that can write and edit books. The quick answer: Yes.

## The fossil record

*Fossils* are the preserved remains of the bodies of dead organisms or the remains of the organism's actions — things such as footprints or burrows. The total of all fossils is called the *fossil record.* The fossil record informs scientists about evolution in several important ways:

✔ In the past, creatures that we don't find today lived on the planet.

✔ Not all creatures alive today are represented in the past.

✔ Through time, the physical complexity of organisms has increased. The earliest organisms that scientists can identify were single celled; now complex creatures exist.

✔ The earliest forms of life were aquatic; terrestrial forms appeared later.

The fossil record, incomplete though it may be, is a record of change through time. This record gives us clues to the progression of the development of life on Earth: Small single-celled organisms evolved into more complex ones; life started in the oceans and only later moved onto dry land. The fossil record provides a rough draft of the tree of life. (Head to Chapter 9 for a detailed explanation of the role that the tree of life — code word *phyolgenetics* — plays in evolution.)

A few things about the fossil record stymied Darwin and others in his day, however:

✔ They seemed to find a lot of older rock that had no fossils and newer rock that had complex life forms, making it seem as though complex life forms appeared suddenly.

✔ It wasn't clear why certain fossils were found in the locations where they were found — marine fossils on mountaintops, for example.

✔ Darwin was puzzled by the sudden changes from one type of fossil to the next, when there seemed to be very few, if any, transitional life forms.

The following sections explain what modern science says about these issues.

### *Conundrum 1: The seemingly sudden appearance of complex life forms*

In studying the fossil record, scientists in Darwin's day were limited in a couple of ways that scientists today aren't:

✔ Their fossil record was even more incomplete than ours today. They didn't have any fossils older than 500 million years.

✔ They lacked the technology to find microscopic fossils. What Darwin and others perceived as gaps in the fossil record actually weren't gaps at all. Today, scientists have the advantage of a much more thorough search of the planet for older fossils and, more important, far more sophisticated techniques for identifying microscopic fossils in rocks.

The earliest fossils that scientists find are single-celled organisms, which Darwin lacked the ability to see physically. Today, scientists know that life has existed continuously on Earth for about the past 3.5 billion to 4 billion years, and they see the same increase in complexity through time that scientists observed in Darwin's day.

# Fossils: Not just rock anymore

Back in Darwin's day, everyone knew that what was so cool about fossils was that biological material had been turned to stone through a process of mineralization. But today, people know something even cooler: Some of the biological material can survive this process and persist for a very long time. Scientists have been able to isolate DNA from organisms, like mammoths and cave bears, that died tens of thousands of years ago.

In retrospect, this feat is not as surprising as it first sounds. DNA is awfully tough stuff; your survival depends on it, after all. Also, techniques for isolating DNA are becoming more and more precise, allowing scientists to work with smaller and smaller quantities.

Even more amazing, scientists recently showed that soft tissue can survive for at least 68 million years inside fossilized *Tyrannosaurus rex* bone. It hasn't been possible (yet) to isolate DNA from such material, but it has been possible to determine the amino acid sequences of some of the remaining proteins. Tyrannosaurus rex proteins show considerable similarity to the proteins of modern birds — it turns out that *T. rex* might've tasted a lot like chicken — not surprising, given the close relationship suggested between dinosaurs and birds by the fossils of *Archaeopteryx* and other feathered dinosaurs. Now scientists have biochemical evidence supporting the same connection.

## Conundrum 2: Marine fossils on mountaintops

Today, scientists understand the process of *plate tectonics* — the moving around of large chunks of the Earth's surface. The idea that the continents may not be fixed in place was greeted with skepticism as recently as 50 years ago, yet now we know that Earth's crust is composed of a series of plates that move relative to one another, fuse, and break apart, resulting in earthquakes and volcanoes.

What seemed like science fiction a little while ago is now something that science can routinely observe and measure. Submarines can dive to the depths of the ocean where plates are separating so researchers can measure the process. Very accurate markers can be placed in different locations across fault lines and their relative movements can be tracked with satellites and lasers. Scientists know, for example, the rate at which parts of California are moving apart and mountain ranges are pushing higher.

The fact that continents move explains why fossils turn up in the unlikeliest places: tropical fossils in Antarctica, for example (biologists have every reason to believe that Antarctica was once in the tropics), or seashell fossils on mountaintops (rocks that were once at sea level can be pushed upward over long periods to form mountain ranges). By understanding more about geological processes and time scales, the fossil record is more comprehensible.

### Conundrum 3: The seeming lack of transitional forms

Darwin wondered where to find the transitional life forms (consider them the in-between-this-and-that forms). Although we've had more success in finding transitional life forms, today's scientists feel his pain. They are better at knowing where to look and they have more people looking, but they still struggle to find them.

Scientists hypothesize that evolution doesn't occur at a constant rate: It can occur in bursts separated by long periods when not much happens. If the transitional period was brief, the chance that such forms would have been fossilized is even more dicey.

When you think about it, it's really quite amazing that people find fossils at all because the events leading to the fossilization of those wonderfully complete skeletons you read about in the news are quite rare. Conditions have to be just right — leave something in your vegetable crisper for sixty million years, and it's not likely to be enough material to do any sort of analysis on! The organism not only has to die, but it also has to be buried intact and remain undisturbed in conditions hospitable to the mineralization process that preserves the remains. Then, possibly millions of years later, someone stumbles across it and calls in the news cameras.

## Biogeographic patterns, or location, location, location

Darwin carefully studied the biogeographical patterns of existing species. (*Biogeography* is the study of the locations of different species through space.) Biogeography reveals that species that appear to be closely related tend to be geographically close as well, as though groups of species had a common origin at a particular geographic location and radiated out from there. Darwin was especially interested in the study of species on islands, and he observed that they seemed to be most closely related to species found on the nearest mainland. Darwin was interested in what, if anything, these biogeographical patterns revealed about evolutionary history.

In developing his ideas, Darwin focused on finches that lived on the Galapagos Islands, an archipelago in the Pacific Ocean off South America. Several species of finches live on the Galapagos, each species inhabiting a different island. The species seemed quite similar to one another *and* to a species on the mainland, leading Darwin to hypothesize that the different species of Galapagos finches were descended from individuals in the mainland species that had reached the islands sometime in the past. Because conditions on the islands differed from conditions on the mainland, the selective pressures acting on the finches also differed, resulting in new traits being favored in the new environment.

## Archaeopteryx and Tiktaalik

*Archaeopteryx* is one of the few transitional life forms that was known in Darwin's time. This early birdlike creature had many characteristics in common with some dinosaurs, yet it also had wings and feathers. Most obvious to the casual observer, *Archaeopteryx* had jaws full of sharp teeth, rather than the beak structure of birds. *Archaeopteryx* was clearly more toward the bird end of the transition to flight. Recently, paleontologists have discovered feathered dinosaurs that did not have wings.

Another interesting creature, *Tiktaalik,* had a skeletal structure intermediate between fish and *tetrapods* (critters with four legs) and had both gills and lungs. This skeletal structure was sufficient to have allowed the organism to support itself, at least briefly, out of water. When the first creatures crawled onto the land, they might have looked like *Tiktaalik.* You can read more about these and other fossil finds in Chapter 20.

As a result of the different evolutionary tracks between the mainland finches and island finches, the gradual changes accumulated to the point where the island finches were different enough from the mainland finches to be considered a new species. The concept of *speciation,* in which one species gives rise to a new species, can seem a bit slippery at first, but it's really not. Head to Chapter 8 for the details.

This process occurred on the various Galapagos Islands, which are far enough apart that travel among them by finches is uncommon. After those rare events when finches *did* make it to a new island — perhaps as a result of being blown there by a storm — they evolved separately from the population on the island from which they came, in response to whatever novel environmental factors were present in their new home.

When Darwin proposed that the Galapagos Islands had become inhabited by so many different but apparently related species of finches through the process of evolution, he had only his observations of existing variation to rely on. Today's scientists, by analyzing DNA, have confirmed these relationships, and detailed studies support Darwin's hypotheses about the existence of different selective pressures on different islands.

## *Natural selection and speciation*

As the preceding section explains, Darwin hypothesized that natural selection operating over a long period accumulates enough small changes in a population to make that population so different that it would be its own species, no longer able to interbreed with other populations of the species it had previously been a member of. Once again, Darwin turns out to have been right. Scientists have evidence that such small changes can have such large consequences over time.

What constitutes a species? The answer depends on the type of organisms you're talking about. Scientists have a reasonably good handle on what constitutes an animal species; determining what differentiates plant and microbial species (such as viruses and bacteria) is a bit slipperier. For animals, though, differentiating one species from another is fairly clear cut. A *species* is a group of organisms that can breed with one another but not with organisms in different species. In other words, reproductive isolation is the key to differentiating species.

Given the way evolution works (small changes over time produce enough changes to create a different species), researchers should be able to find all intermediate forms in nature. *Ring species,* species in which two populations of a particular species can't interbreed with each other (usually because of geographical distance) even though both can breed with other populations of that species, allow scientists to observe how the gradual changes can result in reproductive isolation. In addition to the ring species, scientists have numerous examples of intermediary species: those that have recently diverged and are very similar to one another yet unable to interbreed, and those that are in the process of diverging, in which case they've already differentiated to the point where reproduction is less successful or less common. You can find out more about these patterns in Chapter 8.

# Origin of life

This book concerns itself with the evolution of organisms that are already present, not with the question — fascinating as it is — of where organisms came from in the first place. The question of how life arose on Earth really isn't a question for evolutionary biology. It's a question for chemistry, because in asking about the transition from nonliving to living systems, you must ask questions about the chemical environment that existed on Earth at the time when life appeared. Although no one has succeeded yet with an experiment that involves mixing a bunch of things in a beaker and waiting for something to crawl out, some very clever experiments have been conducted that show how complex biochemicals can arise spontaneously out of simple mixtures under conditions thought to be present on Earth more than 3 billion years ago.

Darwin imagined that such things might happen in a warm pond. Chemists Harold C. Urey (who won the Nobel Prize for discovering heavy water) and Stanley L. Miller actually made the pond. They combined water hydrogen, methane, and ammonia in a sterile glass system; heated the flask to produce a humid atmosphere; and then sent electrical shocks though the mixture to simulate lightning. They repeated this procedure for a week and then analyzed the contents of the flask. By using this simple procedure, they were able to produce DNA, RNA, amino acids, sugars, and lipids — all the building blocks of life from four very simple molecules.

# Chapter 3

# Getting into Your Genes: (Very) Basic Genetics

The study of genetics is pretty fascinating, and the language of genetics is cropping up in a growing number of places. It's often in the news and even makes periodic appearances in movies and TV shows. You've probably heard references to DNA and genes; perhaps you've even heard of the Human Genome Project or of specific genes for human diseases. All this aside, you may be wondering why I've included a chapter on genetics in a book about evolution. The answer is simple. Evolution involves genetic changes over time, so to understand evolution, you need to know a little bit about genetics: what it is, how it works, and what parts are particularly important to the study of evolution.

## What Is Genetics?

*Genetics* is the science of investigating the relationship between parental and offspring characteristics; in other words, it's the study of heredity. For most of human history, people have understood that offspring tend to resemble their parents. Only relatively recently, beginning in the late 1800s, have we begun to understand how the genetic process works.

An easy way to think about genetics is to think about it in terms of information. In every cell of your body is a complete instruction manual for making a person, and somehow, these instructions get passed on to your offspring. The instruction manual is your DNA — basically a repository for all the instructions that make you, you.

As our understanding of the underlying mechanisms of genetics has increased, the field of genetics has expanded to include several new areas:

- **Molecular genetics** is concerned with the biochemistry of DNA and genes, helping scientists understand exactly how DNA is replicated and transcribed. Molecular genetics is important to evolution because it helps clarify the process of *mutation* — that is, the errors that occur when something in the replication process goes awry. Most of these mutations are bad, but every so often, one of them results in something good. Mutations are the initial sources of the variations on which natural selection can act.

- **Genomics,** a new branch of genetics, is concerned with the properties and information content of whole genomes. Comparing the genomes of different organisms gives us a better idea of how, for example, humans can be so different from chimps when they have most of the same DNA sequences. Looking at major genome-wide differences between people can help us understand the health implications of these differences.

- **Population genetics** is the study of how the genetic variation that exists within groups of organisms changes over time. By studying large groups rather than individuals, scientists can observe the evolutionary process — some genes become more common, and others go extinct — to determine whether natural selection is involved.

Population geneticists can't use elephants or any other creature that has a long lifespan. Instead, they perform these studies with critters that don't live as long, such as bacteria. Sure, it takes some of the glamour out of the headlines, but the findings are still pretty amazing (to geneticists, anyway).

It's not always necessary to understand a process completely to make use of it. Although humans have only recently come to understand the details of how heredity works, we've been selectively breeding agriculturally important plant and animal types for thousands of years. For his part, Charles Darwin didn't understand exactly how it was that offspring resemble their parents; he just knew that they did. In the short span of this chapter, you'll discover more about genetics than Darwin ever knew! Because we now have a better understanding of the nuts and bolts of genetics and heredity, we are able to understand the evolutionary process in ways that Darwin could only dream of.

# DNA: A Molecule for Storing Genetic Information

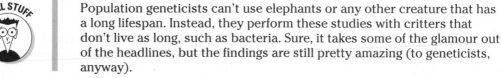

DNA, otherwise known as *deoxyribonucleic acid,* is the molecule that stores your genetic information. As you might imagine, the instruction manual for

making you (or any other type of critter) can be very large, so it's not too surprising that a single molecule of DNA can be enormous — so big, in fact, that
under the right circumstances you can see a DNA molecule with the naked eye.
(To find out how, see the sidebar "A DNA cocktail: Extracting DNA at home.")

# Chromosomes: Where your DNA is

An organism's DNA is found in a cellular structure called a *chromosome*.
Some organisms have all their DNA on one chromosome, while other organisms have their DNA spread across several chromosomes. The DNA of sexually reproducing organisms, like animals and humans, is arranged on pairs of
chromosomes. When these organisms make offspring, the offspring get one
set of chromosomes from each parent. Humans, for example, have 23 pairs of
chromosomes. Each of us got a set from each parent, meaning we each have
half of Mom's genes and half of Dad's.

DNA isn't an abstract concept. It's an actual thing that appears in a particular
place (the chromosomes) in each of your cells. And every cell has a copy of
the chromosomes.

# DNA's four-letter alphabet

Although DNA can be a huge molecule, it's actually a simple one. DNA is
made up of just four different building blocks that are called *nucleotides*, or
bases. The four nucleotides are

- Adenine, abbreviated as A
- Cytosine, abbreviated as C
- Thymine, abbreviated as T
- Guanine, abbreviated as G

The function of DNA is information storage, which is what's so cool about it.
All the instructions needed to make you can be written with just four letters!

The structure of DNA is a double helix consisting of two strands winding
around each other (see Figure 3-1). Each strand can contain up to thousands
of the four nucleotides, and the two strands are joined in a very specific way:

- A always pairs with T.
- C always pairs with G.
- T always pairs with A.
- G always pairs with C.

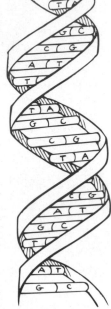

**Figure 3-1:**
The double-
helix
structure
of DNA.

One strand is an inverted version of the other, so if you know the sequence of one strand, you know the sequence of both strands. To give you an oversimplified example, if you have one strand of CATG, you know that the other strand is GTAC.

You may wonder how four nucleotides could possibly be the basis of all life in all its complexity. Well, you wouldn't be the first. What scientists discovered is that these four letters actually appear in groups of three, called *codons*. It turns out there are 64 codons. (You can read more about codons and the genetic code in the section "Protein-coding RNA and the genetic code" later in this chapter).

The double-helix structure of DNA has two properties that make it an excellent molecule for information storage:

- ✔ **It's an incredibly tough molecule.** DNA is so tough, in fact, that scientists have been able to isolate intact DNA from extinct mammoths found buried in Siberian ice and even, in some cases, from fossilized bones.

- ✔ **The double-stranded arrangement provides a very easy way to make accurate copies.** During DNA replication, the two strands of the double helix separate, and an *enzyme* (a protein that's involved in facilitating a chemical reaction) called *DNA polymerase* makes two new strands by using the original strands as guides. This produces two double helix molecules. Each is an exact copy of the original.

## A DNA cocktail: Extracting DNA at home

DNA tends to strike people as being somewhat mysterious, but really, it's just a very long molecule. Although it's certainly amazing how organisms use DNA for information storage, it's not magic — just chemistry. If you want to *see* DNA rather than just read about it, the following recipe for a DNA cocktail shows you how to do so. The chemistry is simple.

**Ingredients**

8 strawberries

⅓ ounce blue Curaçao liqueur

⅔ ounce gin

2 ounces fresh pineapple juice

**Instructions:**

1. **Freeze the strawberries.**

2. **Chill the Curaçao, the gin, and a glass.**

   A tall narrow glass works best. A test tube would work too, but make sure it's a clean one!

3. **When the glass (or test tube) is cold, add the Curaçao.**

4. **Tilt the glass or tube with great care; then pour the chilled gin down the side to form a layer above the Curaçao.**

5. **Purée the frozen strawberries with the pineapple juice for 10 seconds.**

   Strawberries contain DNA. Blending them with pineapple juice allows the enzymatic activity in the juice to free the DNA from all the other bits of the strawberry that it hangs onto. The fresher the pineapple juice, the more enzymatic activity it will have, allowing the experiment to work even better.

6. **Layer the strawberry-pineapple mixture on top of the gin.**

   When the now-dissolved DNA comes into contact with the cold gin, it *precipitates out of solution* (that is, turns into a solid due to the chemical reaction), and you see little white wisps floating in the gin layer. Those white wisps are the actual DNA molecules, and they contain all the information that makes the strawberry plant what it is. The Curaçao doesn't serve any high-tech chemical function; it's just there to make the final product a nice red, white, and blue — and to make it taste nice!

# Reading the Instructions: From DNA to RNA to Proteins

The four-letter alphabet of DNA is where an organism's instruction manual is stored, but when it comes time to actually make an organism, you need proteins. Proteins comprise most of the basic machinery that makes an organism work. Your muscles are made of proteins, your antibodies are proteins, your digestive enzymes are proteins — you just wouldn't be who you are without proteins.

To make proteins, you need RNA. RNA, or *ribonucleic acid,* is similar to DNA but has a couple of important differences:

- RNA is usually a single-stranded molecule, whereas DNA is a double-stranded molecule.

- Although both RNA and DNA are made of four nucleotides, RNA uses uracil instead of thymine. Specifically, RNA uses the nucleotides adenine, uracil, cytosine, and guanine, which are abbreviated A, U, C, and G.

## Transcription: Producing RNA

The process in which a single-stranded RNA molecule is produced from a double-stranded DNA molecule is called *transcription.* The details of the transcription process aren't important for the purposes of this discussion (you can thank me later for sparing you), but the gist of it is that the double-stranded DNA molecule unwinds a bit, and a single-stranded RNA molecule is produced by copying one of the strands.

The four nucleotides that make up RNA line up to match the order of the DNA nucleotides, as follows:

- A (from RNA) lines up with T (from DNA).
- C (from RNA) lines up with G (from DNA).
- G (from RNA) lines up with C (from DNA).
- U (from RNA) lines up with A (from DNA).

In this way, the four-letter alphabet of DNA is transcribed into the four-letter alphabet of RNA.

 Different regions of DNA produce RNA transcripts with different functions. Some of the RNA transcripts code directly for the production of proteins, and other types of RNA transcripts don't code for proteins themselves, they just help the process along.

## Protein-coding RNA and the genetic code

The RNA transcripts that code for the production of proteins are called *messenger RNA* (abbreviated mRNA). Messenger RNA's job is to transfer the information regarding what proteins need to be made from the DNA to the protein-producing machinery where proteins are assembled: the ribosomes (see the next section, "Non-protein-coding RNA").

RNA and DNA are each composed of four subunits (their four respective nucleotides). The DNA nucleotides appear in groups of three (the 64 codons, discussed in "DNA: A Molecule for Storing Genetic Information" earlier in this chapter). Proteins are composed of 20 subunits, called *amino acids*. The exact translation from codons to the amino acids is called the *genetic code*.

Back before anyone knew that DNA is the molecule where genetic information is stored, many people argued that DNA just wasn't complicated enough. They wondered how something with only four different letters could code for all the complexity of an organism. Thus, at one point, proteins were considered to be good candidates for the material used for information storage because they had 20 amino acids — or 20 different letters available in their alphabet. The breakthrough in understanding came when scientists determined that the 4 nucleotides were read in groups of 3, meaning that instead of containing 4 individual letters, DNA has an alphabet of 64 triplet letters, called codons. That discovery raised another question. Instead of wondering how to code for 20 amino acids with only 4 letters, people questioned how you would code for only 20 amino acids when you have 64 codons. The answer is that there is some redundancy in the translation of DNA into proteins.

Figure 3-2 shows all 64 codons of the genetic code. Most of the amino acids correspond to multiple codons, as you can see from the figure (the notations Phe, Leu, and so on). Some codons, though, don't code for amino acids at all; instead, they signal that protein synthesis should stop. These codons are called *stop codons*. They tell the protein-producing machinery to stop adding amino acids to a growing protein, and their presence indicates that the protein is finished.

Second Letter

| First Letter | | U | C | A | G | Third Letter |
|---|---|---|---|---|---|---|
| **U** | | UUU UUC } Phe<br>UUA UUG } Leu | UCU UCC UCA UCG } Ser | UAU UAC } Tyr<br>UAA Stop<br>UAG Stop | UGU UGC } Cys<br>UGA Stop<br>UGG Trp | U<br>C<br>A<br>G |
| **C** | | CUU CUC CUA CUG } Leu | CCU CCC CCA CCG } Pro | CAU CAC } His<br>CAA CAG } Gln | CGU CGC CGA CGG } Arg | U<br>C<br>A<br>G |
| **A** | | AUU AUC AUA } Ile<br>AUG Met | ACU ACC ACA ACG } Thr | AAU AAC } Asn<br>AAA AAG } Lys | AGU AGC } Ser<br>AGA AGG } Arg | U<br>C<br>A<br>G |
| **G** | | GUU GUC GUA GUG } Val | GCU GCC GCA GCG } Ala | GAU GAC } Asp<br>GAA GAG } Glu | GGU GGC GGA GGG } Gly | U<br>C<br>A<br>G |

**Figure 3-2:**
The 64 codons of the genetic code.

# Non-protein-coding RNA

In addition to the protein-coding transcripts, an organism's DNA produces other RNA transcripts that assist with the production of proteins. These other RNAs fall into several categories. For purposes of this book, I won't bore you by listing them. All you need to remember is that the most important category for evolution is ribosomal RNA.

*Ribosomal RNA* (abbreviated rRNA) is the class of RNA molecules that makes up the *ribosomes,* the cellular factories that produce proteins.

Ribosomal RNA is of special interest to evolutionary biologists because all organisms need ribosomes for protein production; therefore, ribosomal RNA can be used for a couple key evolutionary tasks:

- ✔ **Understanding relationships between different organisms:** Ribosomes provide a character that can be compared across all branches of life. Most characteristics aren't shared across all branches of life. Take eyes, for example. Few things have eyes. As a result, comparing a human with, say, a stalk of broccoli and a mushroom based on similarities and differences in the structure of the eye is impossible. But humans, mushrooms, and stalks of broccoli *do* all have ribosomal RNA. (In case you're curious, you and the mushroom are a fair bit more similar to each other at the ribosomal level than either of you is to the broccoli!)

- ✔ **Determining historical relationships between species:** Beyond explaining the level at which people and fungi share similarities, the really fascinating thing about ribosomal RNA is that evolutionists can use nucleotide sequences to determine the historical relationships among species. In other words, it can help clarify which branch of the tree of life an organism belongs to. You can read more about the tree of life in Chapter 9.

Ribosomal RNA isn't important just as a tool for evolutionary biology. It's also important to the proper functioning of the cell — so much so that 80 percent of the RNA in a given cell can be ribosomal RNA. The other categories of non-coding RNA are *transfer RNA* (abbreviated tRNA), which is involved in assembling the amino acids that make a protein, and a growing collection of small RNA molecules (we keep discovering new ones) that seem to be involved with the regulation of *gene expression* — a fancy way of referring to the process of deciding which genes get turned on and off in any given cell. (Think about it: All your cells have all your genes, but the genes needed to make your eyes aren't expressed in your fingers, or vice versa.)

# Getting Specific about Genes

Oh, the poor gene. It has fallen mightily from its halcyon days of being defined as the fundamental unit of heredity. Back then, it was clear that somehow, some way, specific bits of information were passed from parent to offspring. No one had any idea exactly what these bits were, but it was clear that they existed, and we called them genes. Now that we understand genetic information is contained in an organism's DNA, we've booted . . . er, defined what we mean by *gene* more precisely.

## Of alleles and loci

When most people think about heritable traits, they think about genes. Unfortunately, the term *gene* is a little too general for a discussion of evolution. Instead, you need to know a little more about how the DNA strand is put together. As I state earlier in this chapter, DNA is simply long strings of nucleotides, abbreviated A, C, T, and G. That's good for starters, but you need to know a bit more.

The word *gene* is commonly used in a couple of different but related ways:

✔ **To refer to a specific place in an organism's genome:** An organism's DNA contains a *lot* of information, including all the instructions for building and maintaining that particular life form. The different instructions for different genes are located in different places along the DNA sequence. A specific location of a gene in an organism's DNA is called a *locus* (plural *loci,* pronounced *low-sigh*).

✔ **To refer to the exact sequence of the DNA at a specific place (or locus) in the genome:** The sequence of nucleotides at a specific locus can differ among organisms. These different sequences are called *alleles.* Take the locus that stores the information to make the blood type protein. When scientists examine this locus in several individuals, they find variations; some alleles code for type O blood, some for type A blood, and so on. The exact sequence of the ACTG alphabet is different, and this is why people have different blood types.

Using the first, you might refer to a gene for "blood type," but using the second, you'd refer to a gene for "type O blood" — the specific DNA sequence that results in a particular blood type. Both definitions are workable as long as you keep track of which one you're using. In this book, when I talk about genes, I'm referring in most instances to the *loci.* And when I'm referring to alleles, I let you know.

At any given place *(locus)* on the DNA strand, the sequence of nucleotides (the alleles) that appears can be the same or different, with each particular sequence representing a different manifestation of a particular trait. At the locus where the eye-color trait resides, for example, you find alleles representing the different colors: blue, green, brown, and so on.

Imagine a single locus with two alleles, A and a, each representing a different manifestation of a particular trait. Because the locus has only two options, you can figure out pretty easily what individual combinations you may find in the population. In this case, some people would have AA (having received A from both parents), Aa (having received A from one parent and a from the other), and aa (having received a from both parents).

Here are a couple of other facts about alleles to keep in mind:

✔ As is the case with eye color (or blood type or a number of other traits, for that matter), you can have more than two alleles at a given locus *in the population,* but *any particular individual* can have at most only two alleles: one that came from the mother and one that came from the father.

✔ When two alleles at a particular locus are the same, they're said to be *homozygous.* When the alleles are different, they're *heterozygous.* I don't throw these terms in just for kicks. Whether the alleles are different or the same is an important factor in how, when, or even whether the heritable trait manifests itself, as the following sections explain.

## Dominant, recessive, or passive-aggressive?

You probably won't be surprised that the genes you (or any other sexually reproducing organism) inherited from your parents can determine outward characteristics — hence the idiom "chip off the old block." The scientific way of stating this rather obvious point is "Genotype (genes) translates into phenotype (physical traits)." Here are the possibilities:

✔ **Each different genotype has different phenotypes.** Maybe AA makes red flowers, aa makes white flowers, and Aa makes pink flowers — an easy-to-understand example if you know that the A allele codes for a red-pigment protein, and the more of that protein the individual makes, the redder the pigment is.

✔ **The genes interact as dominant (represented by a capital letter) and recessive (represented by a small letter).** When a recessive allele is paired with a dominant allele, the phenotype is the same as for an individual with two dominant alleles. So in the AA, Aa, and aa example, AA and Aa have one phenotype, and aa has another.

To read more about genotype and phenotype — two key terms in evolutionary biology — go the later section "Genotype and phenotype."

# Summing It All Up: Genomes

The sum of all the DNA in an organism is called the organism's *genome*. Studying the genome can reveal quite a bit. Scientists know, for example, that a gene exists for a trait like eye color. But sequence an entire genome, and they see that a lot of genes whose purpose isn't clear. They scratch their heads and say "I wonder what this one does?" or "Wow, the sequence of that gene looks a lot like a particular fish gene whose function we do know." Having a whole genome lets scientists asked questions about genes in a completely different way.

From an evolutionary point of view, the genome is intriguing because it presents evolutionary scientists with a bunch of deep questions to ponder like, if genome sizes are different for different organisms (and some in ways that make no obvious sense), then how did that happen? Or why does so much of the genome seem to be junk (and why is this the case in some species but not others)? Any why are similar genes in different places on the genomes of different species?

## Size isn't everything: Sizing up the genome

Different organisms have radically different genome sizes, but not necessarily in the way you might expect. You have a much bigger genome than a bacterium or a mushroom does, and at first, that seems to make a lot of sense. After all, humans certainly appear to be more complex than bacteria or mushrooms. We have a lot more parts — arms, eyes, complicated nervous systems, and so on — so it seems reasonable that our genome would be bigger. Right? Well . . . maybe, and maybe not.

The range of genome sizes varies among several major groups of organisms. Your genome is much bigger than the genome of yeast or *Escherichia coli* (*E. coli*) bacteria, for example, yet many plants have more DNA that people do. And just in case your ego hasn't taken enough of a hit, you should be aware that some plants don't just have more DNA than we do, but more genes as well. Poplars, for example, have twice as many genes as people do.

For details about genome size and coding and non-coding DNA, head to Chapter 15.

### Base pairs

The unit of genome size is the number of *base pairs* (abbreviated *bp*). Why base pairs? It's simple math, really — and understanding how the DNA sequence works. Think about the size of the human genome, which contains about 12 billion nucleotides. Because of the paired structure of DNA (A always pairs with T, C always pairs with G, and so on), when you know the 6 billion on one strand, you automatically know the 6 billion on the other strand. From an information standpoint, only 6 billion independent bits of information exist. As a result, scientists refer to the size of the human genome as 6 billion bp.

### Junk DNA

Some DNA doesn't seem to contain much information; it's what scientists call *junk DNA*. Some junk DNA consists of long sections in which short sequences of nucleotides repeat over and over and over. Both plants and mammals have lots of junk DNA, but plants seem to have more (which explains why the fern hanging in your kitchen has more DNA than you do). That's quite a feat, considering that we humans have a lot of junk ourselves. And just how much junk do we have? As a matter of fact, it appears that more than 95 percent of human DNA doesn't contain any information about how to make a person! Researchers are still trying to figure out why so much of human DNA is junk — a problem that I discuss in Chapter 15.

# Number of genes

Scientists have determined the DNA sequence of the entire human genome (a fact you may already know from the various news stories that accompanied the completion of the Human Genome Project). So it may come as a surprise that scientists still don't know exactly how many genes humans have. Are they lazy or just unmotivated? As it turns out, neither.

It's sometimes tricky to tell exactly when a sequence at some particular locus in human genome contains instructions used for making a person and when it contains junk. And because the human genome is more than 95 percent junk, small variations in how scientists go about deciding what is a gene and what isn't can make a big difference in the total number of genes they think humans have.

As researchers keep refining their techniques for identifying genes in this big sea of DNA, they end up revising their numbers progressively downward. Currently, they think they have a good handle on the situation and are reasonably certain that the human genome contains about 25,000 genes. Is that a lot? "A lot" is a relative thing. Common intestinal bacteria have about 5,000 genes; yeast has about 6,000; the common laboratory roundworm has around 18,000; and the fruit fly *Drosophila* has 14,000. And many plants seem to have as many genes as human do, and some have far more.

## What's up with all this junk?

The human genome is not the only one that scientists have sequenced in its entirety. Initial genome sequencing projects concentrated on smaller creatures, such as bacteria, and it turns out that very little of those genomes is junk. Although researchers can't say with certainty why humans have so much junk DNA, it's not hard for them to come up with reasons why smaller, rapidly dividing organisms do not. For example:

✔ They're little, so all that junk DNA won't fit.

✔ They're in a hurry, because competing bacteria are sucking up nutrients and reproducing as fast as they can. They don't have time to replicate an enormous genome that's mostly junk without falling behind and being overrun by all the other bacteria.

Humans, however, are not in such a hurry all the time, and the energy it takes to copy human DNA is a very small part of our total energy budget.

When scientists look at a sequence of DNA and identify genes, they have techniques for determining that a particular piece of DNA makes something and can use the genetic code to determine the amino-acid sequence of the protein that piece of DNA makes. But they still don't have a very good way of looking at a protein sequence and figuring out what the protein actually *does*. If the protein looks like something else whose function scientists understand, they have some clues. But if the protein isn't similar to something scientists already know about, they're often in the dark about what particular genes do. Thus, researchers don't yet understand exactly why organisms have the numbers of genes that they have.

## Genome organization: Nuclear, mitochondrial, or free floating?

Where in the world is your genome, anyway? In two different places within your cells: the nucleus and the mitochondria.

✔ **In the nucleus:** Your body is made of cells that have a structure in the middle called the *nucleus*. This is where most of your genetic material resides. The nucleus isn't just one long piece of DNA, but an arranged series of individual pieces called *chromosomes* (refer to the earlier section "Chromosomes: Where your DNA is" for information on what chromosomes do).

✔ **In the mitochondria:** Another structure in your cells contains DNA. This structure is called *mitochondria,* and each of your cells contains several dozens to hundreds of these structures. Mitochondria serve as the power plants of the cell; they're involved in metabolism and energy production.

Mitochondrial DNA doesn't contain many genes, but it's fascinating that mitochondria have any DNA at all. Why do these small things inside your cells have their own genome, and how did they get there? In later chapters, I get back to mitochondria, examining how they evolved and looking at what various studies of mitochondrial DNA tell scientists about human history. For now, just remember that you have some mitochondria and that your mitochondria have some DNA.

Not all organisms have cells with nuclei. In bacteria, the genetic material is happy to float around inside the cell, hanging out with everything else, and doesn't need its own special home. Furthermore, bacteria tend to keep their genome in one piece. It probably would be biochemically complicated for humans to have their entire genome in one segment, as the DNA would be extremely long. Bacteria can get away with keeping their genome together only because their genomes are much smaller. But small things don't necessarily have genomes that are just one piece; many viruses store their genomes in sections.

## How many copies?

Some organisms have more copies of their genome than others do. At first, you may think it would be good to have extra copies of your own personal blueprints, just in case you lose one of the instructions. Although having an extra copy has obvious advantages, it also may have some costs, including the additional time it takes to replicate two copies of your genome before cell division. Biologists think that these costs must outweigh the benefits for most bacteria, which is why many bacteria have only a single copy of their genome.

Sometimes, when the benefits of having a backup copy outweigh the costs of growing more slowly, an organism has an extra copy of its genome. These cases include organisms that really need multiple spare genomes to fix errors. One such organism is the bacteria species *Deinococcus radiodurans.* This little critter has not one extra copy of its genome, but several copies, probably due to the fact that it lives in extremely harsh environments where DNA damage is more likely to occur from such factors as extreme drying. As a result of having these extra copies of its genetic material, *Deinococcus radiodurans* is the most radiation-resistant organism known. This little fellow can handle 500 times more radiation than you can.

# Passing It On: Sexual Reproduction and the Genome

*Diploid* organisms — those with two sets of genetic information — don't have two copies of one genome. Instead, each locus on the genome is comprised of two sets of alleles.

When a sexually reproducing organism — such as a person — produces off-spring, it first must make gametes (in the case of a human, eggs or sperm). Each of the gametes gets one copy of the DNA segments. This gamete combines with a gamete from the organism's mate to produce an individual with two sets of DNA.

When one set comes from Mom and one set comes from Dad, the sets may be slightly different. This situation makes for some interesting genetic questions. One example: How do our bodies read the instructions if the two copies aren't the same? That depends on the particular alleles: what trait or characteristic they're coding for and whether either is dominant.

## Dominating issues

Consider the snapdragons. Snapdragons are diploid, just like you are, though they make ovules and pollen rather than eggs or sperm. At a particular place in their genome is a locus that's responsible for flower color. Two possible alleles appear at this locus; call these alleles W and R. (Alleles are often referred to by letters.) Plants with two copies of the W allele are white; plants with two copies of the R allele are red. How about plants that have one copy of each allele? In the case of snapdragons, the plants with one W allele and one R allele are pink, which seems to make sense.

But it doesn't always work that way. Sometimes one of the alleles is *dominant,* explained in the earlier section "Dominant, recessive, or passive aggressive." If a plant has only one copy of that allele, that allele determines the organism's characteristics. In the pea plants that the famous geneticist Gregor Mendel worked with, there are two possible alleles at the locus responsible for flower color: one that codes for purple flowers (P) and one that codes for white flowers (w). In Mendel's peas, the purple allele is dominant (hence the capital rather than lowercase letter). If a plant has two alleles that code for purple, then it has purple flowers. If it has two alleles that code for white, then it has white flowers. But because the purple allele is dominant, a plant with one purple and one white allele turns out just as purple as a plant with two purple alleles.

In this case, knowing the color of the organism doesn't give you perfect information about the underlying genetics. If the pea flour is white, you know that it has two of white alleles (ww). But if the pea flour is purple, it may have two purple alleles (PP) or one purple allele and one white allele (Pw). This situation leads to the topic of genotype and phenotype, which very conveniently comes next.

## Genotype and phenotype

*Genotype* refers to the alleles that a particular organism has — the actual sequences of DNA in its genome, such as a gene for growth hormone. *Phenotype* refers to the physical characteristics of the organism, such as the organism's height.

Genotype and phenotype are often connected, but the important thing to remember is that the connection is *not* always absolute. Organisms with the same phenotypes may have different genotypes; similarly, organizations with the same genotypes may have different phenotypes. What this means is that you can't always determine what DNA sequences are at play simply by identifying outward characteristics; neither can you always know whether an organism's characteristics are the result of genetics or of something else.

### Different phenotypes, same genotype

An organism's phenotype — its physical characteristics — is not always determined by its genes alone. Grasping this concept is especially important when you think about evolution by natural selection. Environmental factors — like how much food we get or how big our pond is — interact with genotype to produce an organism's phenotype.

Imagine a pack of cheetahs chasing a gazelle across the African plains. One by one, the cheetahs get tired and give up. But one cheetah keeps at it and eventually catches the gazelle. It's a tough year out there in the Serengeti, and the mere difference of a single gazelle can determine whether a cheetah is able to reproduce.

Why this cheetah captured the gazelle while the others fell away, I have absolutely no idea. Are her genes especially good? Does she have different genes that make her go extra fast? Perhaps . . . but perhaps not. It could be that all the cheetahs have the same genes, but this particular cheetah was lucky enough to have been very well fed when she was a cub and, as a result, grew up to be faster and stronger. If that's the case, all these cheetahs have the same genotype, but the phenotype — in the successful cheetah's case, the strength and stamina to continue the chase — differs.

### Same phenotype, different genotypes

Occasionally, organisms have the same phenotype (characteristics) but different genotypes (gene sequences). Consider the human blood-type alleles A, B, and O. Each of us has two of these alleles, receiving one from each parent. Six pairs of alleles are possible: AA, AO, BB, BO, AB, and OO. Yet only four blood types exist: A, B, AB, and O. How does that work?

Well, in one way, it's similar to the snapdragon with a red, white, and pink flower. Whether you have A, B, or AB blood depends on the genotype:

- The AA genotype (you got an A allele from each parent) gives you type A blood.

- The BB genotype (you got a B allele from each parent) gives you type B blood.

- The AB genotype (you got an A allele from one parent and a B from the other) gives you type AB blood.

To understand the others, you need to know that A and B are dominant over O, but neither is dominant over the other. Therefore:

- The AO genotype gives you type A blood.
- The BO genotype gives you type B blood.
- The OO genotype gives you type O blood.

Knowing the phenotype sometimes gives you complete information about the genotype, such as the phenotypes for type O blood and type AB blood. But in the cases of blood type A and blood type B, two possible genotypes could bring about each phenotype.

## What this has to do with natural selection

Natural selection — the process by which organisms with favorable traits are more likely to reproduce and pass on their genes (see Chapter 5 for more in-depth info) — acts on phenotype, not genotype. Think about it. No matter how great an organism's genes are, they can be reproduced in subsequent generations only if that organism reproduces — and *that* depends on outward characteristics (phenotypes).

In the earlier cheetah example, you expect that faster cheetahs will leave more descendents in future generations than the slow ones that couldn't catch dinner. Some phenotypic variation existed in the cheetah population, and as a result, some cheetahs did better than others. That's natural selection. But is it evolution? Depends.

Selection results in evolution *only* if there are genetic differences between the cheetahs of different speeds. The difference in speed has to be a heritable one — one that can be passed on genetically from parent to offspring. If the faster cheetahs have different genes than the slower cheetahs, then the next generation will have a higher proportion of those genes because the cheetahs that had them ate more antelope and made more baby cheetahs. But if the faster cheetahs aren't any different genetically than the slower ones, then there won't be any evolutionary change from one generation to the next.

# Part II
# How Evolution Works

The 5th Wave                    By Rich Tennant

"As you can see, the mutation exists across species, but specifically to those inhabiting the North Pole."

# In this part . . .

*E*volution by natural selection requires mutations. Now, there aren't many times when you hear the word "mutation" and think it's a good thing. Even in evolutionary terms, most mutations are bad. But some mutations give an organism a fitness advantage resulting in its being able to survive and reproduce. The result? More of the genetically advantaged organisms in future populations.

But selection isn't the only evolutionary force. Genetic drift (a fancy way of saying random events) can affect gene frequencies, too. If a mudslide wipes out a large portion of a wildflower that just happens to bloom pink in a particular area, then there will be fewer pink-flowering plants in later generations.

This part tackles the mechanisms of evolutionary change (variation, mutation, selection — both natural and artificial — genetic drift, and so on); their results (loss of genetic variation, change in genetic variation, and speciation); and how to retrace past evolutionary events and see the relationships among species (through phylogenetic trees).

# Chapter 4

# Variation: A Key Requirement for Evolution

*E*volution can't happen without variation, but not just any type of variation will do. It must be *heritable* variation — that is, variation that gets passed genetically from parent to offspring. But some evolutionary forces, such as natural selection and genetic drift (covered in Chapters 5 and 6, respectively), actually cause a *reduction* in genetic variation over time. Fortunately, *mutations* — random errors in the genetic code — generate the very kind of variation that evolution needs. Despite the fact that most of the specific mutations aren't good for any organisms, mutations in general are absolutely necessary.

Variation and mutation go hand in hand, so this chapter examines both topics. And because it's always fun to see how the genes you get affect who you are, I also discuss how many of the traits you exhibit are determined by the genes you have.

## Understanding Variation

In evolutionary terms, *variation* simply refers to the differences you see among individuals. In other words, individuals from any species aren't all the same. Look around a room full of people, and you notice that all of them look different. That's variation. You can see how heritable changes — hair color, eye color, facial structure, height, and so on — manifest themselves outwardly. But these changes also manifest themselves in ways that aren't so easy to see, such as propensity for certain illnesses or blood type. And if those differences are a result of genes, that's *heritable,* or *genetic, variation.*

Even though variation is a simple concept to grasp, it's a crucial component of evolution. Without heritable variation, evolution couldn't occur, because change couldn't occur from generation to generation. If every individual in one generation were the same, every individual in the next generation would be the same. Variation has to be present before natural selection occurs; you can't have the sorting-out process if nothing's available to sort.

You're most aware of small differences in things you spend a lot of time interacting with. Just because all hyenas (or wild dogs, ducks, or any other nonhuman organism) seem *to you* to look the same, act the same, and do the same things doesn't mean that those species lack variation.

## Key concepts in variation

As stated previously, for evolution to occur, the variation has to be heritable. If all the individuals in a population are genetically identical, and they produce offspring that are genetically identical, evolution hasn't occurred, because the next generation is the same genetically as the current generation.

Variation exists in natural populations, and there's often a lot of it:

- **Variation within a species:** All species have variation. Just thinking of the variations within our own species is a good place to start. It would take at least 100 books this size to list all the ways that people are genetically different from one another. An important thing to remember, though, is that the variation within a species doesn't need to be uniformly distributed across all population in that species. The different populations will all have genetic variability, but they all won't always have the same genetic patterns. For example, a species may contain individuals of many different heights, but not all populations within that species will have individuals of all heights; some populations may be made up of, on average, taller individuals than others.

- **Variation within populations:** It's the variability within a population that allows the population to respond to natural selection. Take the finches in the Galapagos. Within one finch population that was studied, there was existing variation in beak size. When other birds arrived on the island and began eating the food these finches relied on, this existing variation allowed that species to evolve to better utilize a smaller seed resource.

- **Variations between populations:** It's the variation between populations that tells us that two populations have evolved in different ways. The populations are genetically different from one another, meaning that gene flow between the populations hasn't been able to overwhelm the different evolutionary forces they experience. That different alleles have increased or decreased in the different populations could be the result of natural selection (discussed in Chapter 5), random factors (the topic in Chapter 6), of both. If two populations continue to become genetically different, they can become too different to interbreed, a topic addressed in Chapter 8.

In evolution, the relevant variation occurs at the group or species level. You cannot talk about variation unless you have more than one individual. If you have variation in the genetic composition of the group or species, the next generation can be different. *That's* evolution.

A variety of mechanisms can cause the genetic makeup of the next generation to differ from the genetic makeup of the current generation. Natural selection (Chapter 5) and random factors (Chapter 6) are two immediate examples. These mechanisms change the genetic makeup of a population from one generation to the next: natural selection, because some genetic types have better reproductive success than others, and genetic drift, just because sometimes stuff happens — a fire kills many blue-eyed deer, leaving fewer blue-eyed deer to make more blue-eyed deer.

## Two kinds of variation: Phenotypic and genotypic

The two kinds of variation are *phenotypic* (changes in outward physical traits) and *genotypic* (changes in the organism's underlying genetic makeup — basically, the DNA sequence of its genes).

Individuals with the same phenotype can have different genotypes, and individuals with the same genotype can have different phenotypes. Height, for example, has both a genetic component and an environmental component. You'd think that two people with identical genetic makeup (such as identical twins) would be the same height, but if only one received a healthy diet, the result could be two adults of different heights. In this example, two people with the same genotype have different phenotypes. Conversely, two people with different genetic make-ups can be the same height. In this case, the two individuals have different genotypes but the same phenotypes.

Keeping this distinction in mind is important when you think about how natural selection can cause evolutionary change. Because natural selection doesn't "see" the different genotypes, it acts only on the *phenotypes* — the physical features of the organism that can influence its survival and reproduction. But evolution can occur *only* if phenotypic differences are correlated with genotypic differences.

Imagine that being taller carries a selective advantage — that is, taller folks are better at surviving and reproducing. In the case of the identical twins with different heights, the taller of the two identical twins would be more likely to contribute genes to the next generation, but because the height difference was due to environmental factors and not genetic makeup, the genetic composition of the next generation doesn't change. In the case of the two unrelated people of the same height, the two individuals have different genotypes, but because natural selection acts only on phenotypes, both individuals are the same height, so they have an equal chance of contributing genes to the next generation.

## *Variation that's important to evolution*

You can partition variation by whether it's heritable (or not) or has fitness consequences (or not):

- **Heritable variation:** This type of variation is genetic; these are the differences at the DNA level that are passed from parent to offspring.

- **Non-heritable variation:** This type of variation is the result of the environment factors, things like diet, amount of sunlight, and so on — not genes.

- **Variations that have fitness consequences:** These variations impact, for better or worse, how well an organism is able to survive and reproduce.

- **Variations that don't have fitness consequences:** These variations don't impact either positively or negatively an organism's ability to survive and reproduce.

The type of variation determines in what manner (or even whether) it has an impact on the genetic makeup of future generations:

- If a variation has a fitness consequences but the variation isn't heritable, it won't lead to evolution. For example, people with really cute tattoos may be more likely to pass along their genes, but there's no genetic component to getting a tattoo. Their genes are, on average, no different from anyone else's genes. ***Remember:*** Without heritable variation, there is no evolution (or evolution books, for that matter).

- If a variation is heritable but it doesn't cause any fitness differences, then the frequencies of these differences in a population will change only as a result of genetic drift, not natural selection.

- If variation is both heritable and has fitness consequences, well then, evolution by natural selection can occur.

## Population structure and gene flow

Species are not uniformly mixed. All the people in the human population, for example, are completely capable of mating, but we're not totally genetically mixed, hence, the existence of regional heritable differences. The reason for this lack of mixing is remarkably simple: It's a matter of *gene flow,* which means the movement of genes in space.

The reason the genes within the human population (and the populations of many other species) aren't completely mixed is because we're more likely to mate with others who are nearby. Populations that are close together in space are more likely to exchange genes with each other than populations that are farther away.

Gene flow, the idea that genes can move from one population to another, is a key concept in evolution. You move to Sweden, find a nice, blue-eyed, blond spouse and raise a family. Your children are Swedish, but maybe they've got your eyes — that's gene flow.

# Where Variation Comes From: Mutations

The process of evolution (via natural selection, drift, or both) eliminates heritable variation. If not all the alleles make it into the next generation — because some are selected against or disappear due to random forces — heritable variation eventually goes away. That scenario leads to the subject of mutation, which is the ultimate source of genetic variation.

Mutations are changes in an organism's DNA. If all the deer in the forest have only genes to produce normal noses and one deer is born with a blinking red nose (a heritable trait), a mutation must have occurred in one of the gametes (either the sperm or the egg) that led to that red-nosed baby deer.

DNA is an extremely stable molecule; that's what makes it so good for storing your genetic information. But mutations still occur:

- ✔ DNA is tough, but it's not indestructible. It can be damaged, and although your cells try to repair the damage, they don't always get it right, resulting in mutations. Agents that have this damaging effect are referred to as *mutagens.* Examples of mutagens include the ionizing radiation associated with nuclear material, ultraviolet radiation from the sun, and many chemicals.

- ✔ The process of DNA replication that occurs in all cell division — be it in the gametes (egg and sperm) or in *somatic tissue* (everything else) — is another source of mutations. DNA replication is an error-prone process — sometimes the copying isn't exact, and that also results in mutations. (To find out more about DNA replication, refer to Chapter 3.)

## Mutations leading to cancer

When you hear the term *mutation,* the first thing that may spring to mind are mutations that cause diseases such as cancer. Stay out in the sun too long, and ultraviolet radiation bombards your skin cells. This radiation causes changes in your DNA that affect the regulation of cell division, and all of a sudden, one of your cells doesn't play nice with the rest of your body. These types of mutations certainly are important medically, but they don't affect the genetic composition of your offspring.

Cancer and other such mutations can be important in evolution if there are differences in the mutation rate between individuals, for example, with some individuals being much more likely to die of cancer than others and, therefore, less likely to have descendants in the next generation. In such a scenario, the characteristic of having a higher mutation rate would be the character under selection, not the cancer itself.

Sometimes, mutations are selectively advantageous. (Rudolph the Red-Nosed Reindeer did save Christmas, after all.) But most of the time, they're not. Mutations are either selectively neutral (they have no effect on fitness) or deleterious (they decrease fitness). Regardless, without mutation, there'd be no variation on which natural selection can act. Yet a number of forces can reduce variation from generation to generation. Variations that are selectively disadvantageous would be eliminated from the population over time, and the random forces of genetic drift (see Chapter 6) eventually would purge diversity from the population.

## Important mutations

Mutations occur all the time, but only certain mutations are important for evolution: those that are heritable. For single-celled organisms like bacteria, all mutations are heritable. A bacterium reproduces by dividing, and any changes that occur in the organism's DNA before division are passed on to the daughter cells. That's not the case with animals. Most of the cells in the human body, for example, don't contribute directly to the next generation. Only the germ line (eggs and sperm) do, so only mutations in those tissues can be passed to offspring.

## Which comes first — the mutant chicken or the selective agent?

Here's a major stumbling block many people have in understanding how mutations are involved in the process of evolution. Many people think that selectively advantageous mutations occur *in response to* environmental factors that make them advantageous. The thinking goes this way: All these short-necked creatures are running around on the plains in Africa; their main food supply is the vegetation on the ground, but they eat so much, not enough ground vegetation is left to feed the population. What do they do? They evolve to have longer necks so that they can eat the leaves off the trees. Voilà! Giraffes evolve, and the problem is solved. But this scenario is *absolutely not* how evolution works.

Mutations don't occur in response to environmental factors; they already exist in the population. In the presence of some new environmental factor that makes the mutation beneficial, the organism with that mutation is more likely to survive. In the example of the short-necked creatures in Africa, suppose that a few random mutations result in a few long-necked individuals. The long-necked ones and the short-necked ones still eat the rapidly diminishing grass supply, but the long-necked ones have an advantage because, in addition to grass, they can eat the leaves off trees. Being better fed, they are more likely to survive and reproduce.

This principle — that the mutation already exists in the population — is crystal clear when you witness evolution in progress, which is possible with organisms that have short generational time spans. Take a flask of bacteria, for example; put in some penicillin; and you're left with a flask that contains only a few surviving bacteria, all of which are penicillin resistant. What happened? Some of the bacteria in the flask already had slightly different DNA sequences that made them resistant to penicillin. Examples illustrating the evolution of antibiotic resistance come up several times in this book because they so beautifully illustrates how organisms — both simple and complex — evolve.

Variation needs to exist in a population *before* the evolutionary forces can select for it. Once the new selective force (climate change, new disease, and so on) appears, it's too late for any new mutations to be beneficial. Only individuals that already have characteristics that are beneficial in the presence of the new selective force will have an advantage.

## Different kinds of mutations

A mutation is simply any change in the DNA sequence. As you can imagine, more than one type of change is possible. At each position in the DNA sequence, there are one of four possible nucleotides (A, C, T, and G). The simplest change to imagine is a change where the original nucleotide in a particular position is replaced by one of the other three possibilities. An A is changed to either G, T, or C, for example. Mutations of this type involving just a single point in the DNA sequence are called *point mutations*. More complicated changes are also possible. Whole sections of DNA can be removed, moved from one place to the next, or duplicated. The following sections explain these types of mutations in more detail (head to Chapter 15 for information on the significance of gene duplication).

DNA is a long string of four different nucleotides that thread off in groups of three. The different three-base sequences instruct the cell to assemble different amino acids into a protein. Some three-base sequences instruct the cell where to start along the DNA sequence and where to stop. Not all sections of an organism's DNA are used to code for proteins, but for the purposes of this discussion the important sections are those that do code for proteins. (Refer to Chapter 3 for more detail about how the sequence of the DNA is used to code for the specific amino acids needed to make a given protein and Chapter 15 if you want to know more about non-coding DNA.)

### Point mutations

In a *point mutation,* a single nucleotide in the DNA is replaced by some other nucleotide, resulting in a particular three-letter sequence of DNA that's different. Because of the redundancy of the genetic code, point mutations don't always result in a change in the amino acid and, therefore, don't affect the organism's phenotype, making this particular mutation selectively neutral. There is no change in fitness between the original type and the mutant type.

Sometimes, though, a point mutation results in a different amino acid being used in the production of a protein. In this type of mutation, the protein may have a different structure and may behave differently. If the organism's phenotype is changed, this type of mutation may have fitness consequences. Or it may not — there are many examples where an amino acid change results in the production of a slightly different protein but one that works exactly as well. Changes of this sort may have fitness consequences, or they may be selectively neutral.

At other times, a point mutation could replace a three-letter sequence coding for an amino acid with a sequence that starts or stops protein production. Stopping production of a protein when only part of the amino acid sequence has been assembled is likely to result in a protein with a very different structure and is likely to have a negative affect an organism's phenotype. More often than not, making just half of a protein will be less advantageous than making the whole thing.

### Insertion and deletion mutations

Larger changes also can occur in an organism's DNA. Sections of DNA can be lost or inserted into other sections. These sorts of changes are referred to as *deletions* and *insertions,* and although they don't always have a large effect, it's easy to see how they can.

Deleting a section of an organism's instructions set is not likely to be advantageous. Whatever information is eliminated may prove to be extremely important. Obviously the larger the deletion, the larger the potential problem, but even small deletions can cause major effect.

# Preventing bad mutations

Mutations tend to be bad, so (not surprisingly) mechanisms exist within cells to reduce the probability of mutation. Biochemical mechanisms repair damaged DNA; proofreading mechanisms catch errors that occur during DNA replication. (Yes, each and every one of us has spell check built in!)

So if mutations can be fixed, why do they exist? For a few reasons.

### Mistakes happen

No repair or proofreading system is perfect. I hope, for example, that this book has no typos; it's gone through several editing and proofreading checks. But every once in a while, you find a typo in a book; maybe you'll find one in this book. No matter how hard you try, being prefect isn't possible, and the same is true of cellular biochemistry.

### Trade-offs

A trade-off may occur between speed of DNA replication and accuracy of DNA replication. Although mutations tend to be deleterious, slowing down reproduction is also bad for fitness. Genes responsible for a phenotype that reproduces slowly but accurately would be at a disadvantage against genes that generate a phenotype that reproduces more rapidly and is almost as accurate. Each of the individual descendants of the rapid (but sloppy) organism would be more likely to have a couple of extra deleterious mutations, but many more of them would be around, and the overall reproductive success of less error-prone individuals would be lower.

This argument suggests that the mutation rate itself is a character affected by natural selection. And it is. We know from laboratory experiments that mutation rate is a variable heritable character.

It's easy to see how too much mutation would be disadvantageous, so we know that natural selection will keep the mutation rate from getting too high. It's a bit harder to see how natural selection would *favor* any mutation rate, but there may be a couple of reasons: the short-term trade-offs such as the one described earlier between speed and accuracy, and long-term effects whereby lineages with low mutation rates are eventually eliminated by lineages with slightly higher mutation rates because the slightly higher mutation rate results in at least some favorable mutations. Perhaps the best mutation rate is not too much, but not too little. Consider this the Goldilocks principle: You don't want too many or too few mutations, but *just* the right amount:

- ✔ **Some mutations are advantageous.** Mutation is the ultimate source of the variation on which natural selection acts. Natural selection acts to eliminate deleterious mutations and increases the frequency of advantageous ones.

- ✔ **Maybe a too-low mutation rate is deleterious:** Some scientists speculate that over the long run, a very low mutation rate, while advantageous for individuals in the short-term (because most mutations are bad and you don't want to make mutated kids), may be disadvantageous for the species as a whole. Over long time periods, the selective forces acting on a species are likely to change (climate changes, species move to different habitats, and so on). Without sufficient variation, the species would be unable to respond evolutionarily to these challenges. Such a species could be outcompeted by a species with some higher level of mutation.

  This scenario is speculation and involves the tricky subject of selection acting a level other than that of the individual (a topic you can read more about in Chapter 11). Scientists lack a clear understanding of what sets a lower limit for mutation rate — maybe all the biochemistry involved in DNA replication eliminates the possibility of a zero mutation rate — still, the concept is an interesting one.

Bottom line: Mutations tend to be bad, and in the short term, *not* having any would be good relative to having some. Thinking that a mutation rate is a good thing because you could end up with descendants that are more fit is like thinking that sinking all your retirement funds into the lottery is a good idea because you could end up a millionaire. True, you could get that result, but chances are that you'll end up broke instead.

# Gene Frequency and the Hardy-Weinberg Equilibrium

For asexually reproducing creatures (those with just one parent), reproduction is extremely simple: Make a copy of your DNA and divide. If no mutations occur, parent and offspring are identical.

For organisms that have two parents (called *diploid organisms*), reproduction is a bit more complicated. Each individual has two copies of DNA, one from each parent, and will pass on a single copy to each offspring (the other parent contributes the other copy). At each location in the genome, a diploid organism has two alleles, and one or the other will randomly end up in each gamete. When offspring are produced, the relative frequency of different genotypes produced will be a function of the frequency of the different alleles in the population.

Heritable variation is necessary for evolution by natural selection. The pattern of the existing variation (something that can be measured) can tell scientists about whether evolution is occurring.

To understand how allele and genotypic frequencies change under various evolutionary forces, scientists study what happens when none of these forces is at work. Under such conditions, the Hardy-Weinberg equilibrium states that allele frequencies don't change and predicts what the frequency of genotypes should be in a population. This equilibrium states that if you know the frequency of the alleles in a population, you can figure out the frequency of the genotypes in the next generation if (1) mating is random and (2) no evolutionary forces are changing the allele frequencies in the next generation.

To help you understand the Hardy-Weinberg equilibrium, the following information focuses on how allele frequencies correspond to genotype frequencies at a locus where, among all the individuals in the population, there are only two different alleles.

## What's the big idea?

The Hardy-Weinberg equilibrium helps scientists determine when natural selection or genetic drift is at work. If the results deviate from the prediction, you know that one of the following situations has occurred, because any of them would cause a deviation:

✔ **Gene flow:** Gene flow is one of the factors that can lead to a deviation from the Hardy-Weinberg equilibrium. If one genotype rather than another moves into or out of a population, the genotype frequencies can't be predicted from the allele frequencies.

Suppose that all the individuals with the aa genotype fly away. You're left with a population that consists only of AA and Aa individuals, and you can't predict the frequency of genotypes accurately from the frequency of the alleles. What's important is not just that movement of individuals into or out of the population is occurring, but that this movement is related to the genotypes of the individuals.

✔ **Selection:** Imagine that individuals with the aa allele die without reproducing (which would be the case with lethal recessive alleles). In such a case, only AA and Aa individuals would be left in the population. Again, the frequency of genotypes can't be accurately predicted by the frequency of the alleles. A deviation from expectations under Hardy-Weinberg indicates that some evolutionary force is at play.

## Inbreeding: An example of nonrandom mating

*Inbreeding,* which is a special case of nonrandom mating, occurs when individuals mate with relatives. Because related individuals are more likely to have similar genes, they're also more likely to have similar deleterious recessive genes (which is why the offspring of close relatives tend to have reduced fitness). If these individuals mate, chances increase that two individuals with the recessive trait will get together and produce offspring that end up with two copies of the recessive genes and the condition that the recessive genes cause.

It's been estimated that the average human has perhaps a few lethal recessive alleles. But these alleles are rare in the population and express themselves only if a person happens to have children with someone who has the same lethal recessive gene.

✔ **Nonrandom mating:** The Hardy-Weinberg equilibrium assumes that individuals mate randomly. If that's not the case — if AA individuals prefer other AA individuals, for example — after one generation of such mating, the proportion of individuals that is heterozygous will decrease. (Only matings between heterozygous individuals can produce more heterozygous individuals, but not all their offspring will be heterozygotes; some will be AA and aa.) Matings between homozygous individuals always produce homozygous offspring.

✔ **Random events (that is, genetic drift):** When population sizes are small, random events can cause a deviation from Hardy-Weinberg. For example, while each allele in an individual has an equal chance of getting into a gamete, when only a small number of offspring are produced, just by chance one or another allele might be over (or under) represented, leading to yet another deviation from the expected genotype frequencies. Head to Chapter 6 for more information on genetic drift.

## Using the Hardy-Weinberg equilibrium

Scientists use the Hardy-Weinberg equilibrium when they know the proportions of the different alleles in the population and want to predict the proportions of the different genotypes in the population. Here's a simple example using one locus with two alleles.

Suppose that you decide to measure the proportion of the alleles A and a in a population. Call these proportions p and q, where p is the proportion of A alleles and q is the proportion of a alleles. Because of the way proportions work (they represent portions of 100 percent), you know that p + q = 1. If 70 percent of the alleles are A, 30 percent are a, and p = 0.7 and q = 0.3.

So now you've got the proportion of A and a, and you want the proportion of AA, Aa, and aa. Here's how you do it:

$$p^2 + 2pq + q^2$$

Plug in the numbers you got for p and q, and you get

0.49 + 0.42 + 0.09

Translation: If 70 percent of the gametes are A, 49 percent of the offspring will be AA (0.7 x 0.7 = 0.49). If you don't find that result, you know that other forces are at play in this population.

Imagine that the a allele is a lethal recessive gene. Anyone unlucky enough to get two copies of this gene dies, which means that you won't find any individuals in the population that are aa. That result is a deviation from the Hardy-Weinberg equilibrium. In this particular example, this deviation is the result of natural selection selecting against people with the aa allele.

# Chapter 5

# Natural Selection and Adaptations in Action

*E*volution is nothing more — or less — than changes in the relative frequencies of heritable traits in a group of organisms (whether particular populations or whole species) over time. Simple enough.

But what causes traits to change over time? Often, it's *natural selection,* the process whereby some individuals, as a result of possessing specific traits (keener eyes, bigger leaves, etc.) leave more descendants than other individuals that lack these traits. If these advantageous traits are heritable (and thus are passed on to offspring), then over time, as some traits are favored and others are selected against, populations change — they evolve. Eyesight gets keener because the individuals with weaker eyes are not passing the genes for weaker eyes to future generations. Changes that are the result of natural selection are adaptations. In this example, keener eyes is an adaptation, as are bigger leaves.

Natural selection is not a difficult concept, but many people get all confused about it — particularly when it comes to differentiating between the process of selection (how it works) and the results of selection (the adaptations). This chapter helps you sort everything out.

Natural selection is only one of the mechanisms that cause evolution. Another mechanism — genetic drift — is the topic of the next chapter.

# *Natural versus Artificial Selection*

As Darwin was formulating his theory of evolution by natural selection, he was influenced by the vast body of knowledge on the domestication of plants and animals. *Artificial selection* refers to the selective process when humans are acting as a selective agent. Darwin was aware of the power of artificial selection to affect genetic changes in domestic animals over a relatively short period of time. Imagine, he perhaps exclaimed, what sort of changes might occur over the history of life on Earth!

Charles Darwin wrote *On the Origin of Species by Means of Natural Selection, or the Preservation of Favoured Races in the Struggle for Life*. In this book, one of his key insights was to recognize that the struggle for life had winners and losers, and that it would result in changes in populations through time as the winners contributed their genes to the next generation but the losers didn't.

Darwin saw that the process of natural selection, where the environment in which a population lived could impact what genes made it into subsequent generations, was very similar the process used in animal breeding. The key difference is that in husbandry, humans — and not nature — decide who the winners and losers are.

 When nature is the selective agent, the process is called *natural selection*. When humans are the selective agent, the process is called *artificial selection*. We can use artificial selection to examine the process of evolution in the laboratory, and we can observe natural selection occurring in the wild.

 The process of artificial selection isn't exactly identical to what happens in the natural environment because humans can get pretty creative in their animal and plant breeding. A particular breeding endeavor, for example, could require a cocktail of approaches: perhaps a little directional selection, just a touch of genetic drift, and a dash of in-breeding followed by some more selection. The result is that allele frequencies of the domestic population change, but it's not strictly identical to the natural process.

 Evolution by natural selection (or any other mechanism) is a property of groups, not individuals. Have you ever seen that cartoon of the fish that grew legs and crawled up onto the land? If that seemed confusing, it's because it is. No one fish grows legs. It lives and dies with the same fins it always had. But if, in a population of fish, there is heritable variation in fin structure (and remember, there is heritable variation for pretty much everything) and fish with stiffer, stubblier fins leave more offspring, then the population at some future time will have, on average, stiffer fins. If mutations that results in even *stiffer* fins are selectively favored, the process keeps going until you have fish that waddle around in the mud (such fish exist), and up up up we go from there.

The following sections tease apart in more detail some of the different ways that selection can act.

## Directional selection

In *directional selection,* natural selection favors an extreme phenotype — for example, the fastest individuals. If selection for the extreme phenotype continues through time, the population will become faster and faster so long as sufficient heritable variation in speed is present on which natural selection can act.

Think about a cheetah and its prey. All other things being equal, faster is always better, so natural selection acts to increase the speed of both the predator and the prey from one generation to the next. (Some physical limits exist on how fast animals can run, of course, but this process has generated some very fast creatures on the plains of the Serengeti!)

## Stabilizing selection

*Stabilizing selection* (also called *balancing selection*) favors the middle ground — that is, traits that aren't too hot or too cold, but just right, when *just right* means comfortably in the middle between two extremes. Think of this form as the Goldilocks of selection.

In stabilizing selection, the most-fit trait is intermediate with respect to whatever characteristic selection is acting on. An example is the birth weight of babies. Very small babies have a lower probability of survival, but very large babies are more likely to suffer complications during birth. Babies of intermediate size, however, are most likely to survive to reproduce.

# Adaptation: Changes Resulting from Natural Selection

An *adaptation* is a trait that has resulted from evolution by natural selection. In the example of the cheetahs earlier in this chapter, the ability to run faster is an example of an adaptation. Antibiotic resistance (see Chapter 17) is another example of an adaptation. This trait appears in a bacterial population because of random mutations that made some bacteria better able to survive and reproduce in the presence of antibiotics.

Some folks quibble over whether the evolution of antibiotic resistance is truly the result of natural selection since people are hosing down the world with tons of antibiotics and thus exerting selective pressure of their own. Others say that it is natural rather than artificial selection because even though people altered the environment in this instance, the bacteria responded (naturally!) to that change. (Also, antibiotics occur naturally — lots of microbes make them — and bacteria evolve in response to those. Many microorganisms produce antibiotics to inhibit the growth of other microorganisms. Penicillin, for example, is a natural antibiotic that gets its name from the mold that produces it, *Penicillium chrysogenum*.)

If the whole concept of adaptation seems to be ridiculously straightforward, that's because overall it is. The following sections explain why recognizing an adaptation isn't not always easy though.

### Is it an adaptation—or not?

Distinguishing adaptive from non-adaptive traits isn't easy. Identifying an adaptation and the selective force that caused it is pretty clear cut when it comes to something like antibiotic resistance because we've been around to observe the whole process. The situation gets a bit dicier when you talk about adaptations we haven't observed, because in nature, it's not always crystal clear whether a particular trait is an adaptation.

# Acclimatization

Adaptation, the result of evolution by natural selection, works at the species level, not at the individual level. Acclimatization, on the other hand, occurs within individuals.

Suppose that you decide to visit Denver, which is a mile above sea level and has less oxygen than lower-lying areas of the country. On your first day in the city, you decide to go jogging and find the activity much more difficult than usual. After a few days, however, you begin to feel like your old self and can jog just as you used to. Why? Because your body reacted to the lower levels of oxygen by producing additional red blood cells. Although most people in this situation would say that they've adapted to the altitude, they'd be more accurate if they said that they've acclimatized. This kind of distinction is exactly the idiosyncratic type of parsing that scientists (and English teachers) love.

While you won't be doing any adapting on your trip to Denver, there is evidence to suggest that human populations that have lived for long periods at high altitude show genetic changes that could be adaptations to the lower oxygen concentrations at altitude. When you acclimate in Denver, you make more red blood cells which are the same as the red blood cells you had before. The Sherpas living at high altitude in the Himalayas, however, have differences in blood chemistry that result in their red blood cells having higher affinity for oxygen than others.

### There are limits

Some limits exist to what natural selection can accomplish. If natural selection always favors more-fit genes, all species would be on the road to super-organism — forever getting faster, glowier, taller, whatever. And there'd be no limit in numbers of limbs, eyes, hearts, tails, and fins, if having more of these things means being more fit. But that's not what happens.

- ✔ **Some things may just not be physically possible:** Mammals will never run at the speed of sound.

- ✔ **Others may not be biologically possible:** The mammal lineage seems limited to four limbs — variation in limb number is absent.

- ✔ **Some adaptations may preclude others:** A bird can have wings that function like a penguin or a hummingbird, but not both. This idea is explained in more detail in the later section "You can't get there from here: Constraints and trade-offs."

## Exaptation: Selecting for one trait, ending up with another

An *exaptation* is a trait that resulted from selection for something other than the trait's current function. Think about feathers. Scientists know from the fossil record that the earliest creatures with feathers didn't fly. So what was the purpose of feathers on these early flightless animals? Maybe they served as insulation (as feathers today do), or maybe they served some other function. Regardless of their actual purpose, what scientists do know is that the feathers didn't have anything to do with flight — which they can tell from the skeletal structure of the fossils (these creatures didn't have wings!).

So although feathers subsequently evolved to be used in flight, the benefit of flight wasn't the selective force responsible for their origin. It just so happened that feathers — an adaptation selected by some fitness advantage unrelated to flight — subsequently became something that natural selection shaped into a wing.

Occasionally, you hear the term *preadaptation,* which means the same thing as *exaptation.* The problem with preadaptation is that folks tend to misinterpret the word, thinking that it implies premeditation in the process of selection. But selection isn't premeditated, of course, so the preferred term is *exaptation.* Truth to tell, however, most evolutionary biologists are likely to say *preadaptation* instead of *exaptation* when they're sitting around talking science over a few beers. But that's when happens when scientists get sloppy drunk.

One of the beauties of the evolutionary process is the way that existing structures are altered for new uses. For that reason, scientists always need to be cautious with any specific example in which they state that a particular trait evolved for a particular reason. Why? If you're trying to understand the process of natural selection, you need to make sure that the traits you're examining were actually the result of natural selection.

## Chromosomes in action: Linkage and hitchhikers

The human genome (and the genomes of most sexually reproducing organisms) consists of several pieces of DNA called *chromosomes*. Parents make copies of their chromosomes and pass one copy of each chromosome to their offspring (refer to Chapter 3 for more info on chromosomes). How the genome is put together sometimes has consequences for how natural selection results in changes in gene frequencies over generations.

Imagine that loci are on the same chromosome. One codes for hair color (its alleles are A and a); the other codes for eye color (its alleles are B and b). Suppose the parent has the AaBb genotype. This parent could have the allele combination Ab and aB (or the combination ab and AB) on each chromosome. Because the loci are linked, the transmission of one allele determines the transmission of the other.

If the hair color locus and the eye color locus are on different chromosomes, their alleles sort independently; if they're on the same chromosome, they *may* be transmitted together. Note that I say *may*. Here's where things get more complicated.

The chromosomes can break and rejoin, so even if two genes are on the same chromosome, they aren't necessarily inherited together. What that means is that

- Even if the parent had the AB and ab allele combinations on its chromosomes, there's still some possibility of producing aB and Ab gametes — and this possibility is greater the farther apart the two loci are on the chromosome.

- The closer two genes are, the more likely they are to be inherited together. If the eye color gene is right next to the hair color gene, it's more likely that when you passed on these traits to your children, they got whatever combination of alleles were found together on one of your chromosomes.

Now suppose that a beneficial mutation occurs at the eye color locus, and selection acts on that allele. Because the mutation is beneficial, the person who carries it will be more likely to leave descendants; hence, this particular allele will increase through time. As a result of the close link between the eye color gene and the hair color gene, whatever hair color allele happens to be next to it on the chromosome will also increase in frequency, even though it may not have any significance.

Why is this example important? Because if you don't know that a ho-hum allele can increase in frequency simply because it's next to a wham-bang allele, you may misinterpret the increasing frequency of the ho-hum allele as being evidence of strong selection for that allele. In this case, you may think that something about the hair color gene itself is important evolution-wise, when what really happened is that it hitchhiked a ride because of its proximity to the eye color.

In hitchhiking, changes through time in the allele frequencies at one locus can be caused by selection acting on a different locus.

## But wait — not all traits are adaptations

Some traits are obviously adaptive, such as antibiotic resistance in bacteria. Because scientists understand the selective force — humans added the antibiotics — they know that the evolutionary change they observe is a response to that selection. The increase in frequency in the population of antibiotic-resistant bacteria is an adaptation to the presence of antibiotics.

Other traits are obviously not adaptations. An example is a change at the level of the DNA that doesn't result in a change in the way a gene works or is expressed. Traits of this type are *selectively neutral:* They don't change the fitness of the organisms that carry them, so their frequency increases or decreases based solely on random factors (see Chapter 6 for details). And any change caused by random factors rather than selection isn't an adaptation.

### The spandrels of San Marco

Some traits that may seem to be adaptive aren't necessarily adaptations. Maybe the traits exist for reasons that we don't understand — developmental constraints, past events, whatever. The point is that although the traits appear to serve some purposeful function, that function had nothing to do with their evolution. In a paper written in 1979, Stephen Jay Gould and Richard C. Lewontin warn evolutionary biologists not to confuse current purpose with past adaptations. To make this point, they wrote about the spandrels of St. Mark's Cathedral in Venice.

A *spandrel* is a structure connecting the dome to the rest of the roof in a particular type of architecture. And in St. Mark's, the spandrels have been painted so amazingly, you'd think that they exist solely for the purpose of bearing the remarkable images. Wrong! The spandrels are just necessary consequences of how the rest of the structure is put together, meaning their purpose is solely architectural, not artistic. But that doesn't mean they didn't make a decent canvas for the artists who added the pictures later.

Unfortunately, the distinction between adaptive and non-adaptive traits isn't always obvious. Some traits appear to be adaptations but aren't. Why is knowing the difference important? Because you don't want to be fooled. Stating that things are adaptations that actually aren't can lead to errors in the way you interpret data.

## You can't get there from here: Constraints and trade-offs

*Constraints* are problems that natural selection can't seem to solve. As a result of how an organism is put together, some types of variation just aren't expected to appear. All vertebrates, for example, have (at most) four limbs. Think about it: You never see mice with six legs. And based on human understanding of mammalian development, it's pretty unlikely that Pegasus, with his four legs and two wings, would ever evolve. Being a mammal and having six limbs is just not a variant you ever see. So even if having a few extra limbs conferred a fitness advantage, natural selection can't get there because there is no heritable variation for extra limbs.

*Trade-offs,* the balance between fitness benefits and fitness costs, represent another key concept in evolutionary biology. You can think of trade-offs as being the "jack of all trades, master of none" phenomenon. Consider all the different kinds of birds in nature. Why isn't there a superbird with the talons of an eagle, the webbed feet of a duck, and all the other avian parts you can think of rolled up into one bird? Well, you can start to see the problem right away: It would be hard for an eagle to sink its talons into a poor little bunny if the eagle had webbed feet. A pretty obvious trade-off exists between swimming feet and grasping feet. The evolution of life histories is rife with examples of trade-offs; head to Chapter 10 for details.

## Run, Mouse, Run

Throughout this book, I occasionally bring up the cheetah's running speed as an example of an adaptation. I've always thought cheetahs were fascinating creatures and looked forward to seeing them on the Nature Channel. The nice story that I tell you is that there was variation for running speed, that cheetahs that ran faster left more descendents, and that, as a result, the cheetah population got really good at running. It's a good story, but science is more than just a good story.

# The ghost of evolution past

When they're talking about adaptations, scientists have to be careful not to identify as an adaptation to current phenomena a trait that's really an adaptation to past phenomena that are no longer active.

Until fairly recently (not much more than 10,000 years ago), for example, North America and South America were home to a diverse group of large mammals such as gomphotheres (four-tusked, elephant-like creatures) that would have eaten a lot of the local vegetation and may have been potential seed-dispersers. As a result, these large herbivores would have been a key factor in the evolution of the local vegetation — that is, the plants would have evolved in concert with the herbivores that ate them and dispersed their seeds. Probably, the plants evolved defenses against herbivory (the eating of plants) and also evolved traits that led to increased seed dispersal. (This idea that two organisms evolve in response to each other is called *co-evolution*. You can read about it in Chapter 13.)

One of the main reasons plants produce fruit is to get their seeds dispersed. Animals eat the fruit; the seeds pass out of the animals' guts and land in new environments where they germinate and produce new plants (if the environment is suitable). One obvious benefit of such a system is that the seed ends up in a pile of fertilizer. A disadvantage is that the animal may eat the seed as well. So seeds need adaptations to prevent them from being digested on their journey through the digestive tract — hence, the hard coatings that many seeds have. Yet a coating hard enough to pass through an animal unscathed may actually make it more difficult for the seed to break through the shell and germinate. A seed with a coating that allows it to survive passage through an animal's gut may now need to go through the gut before it can germinate. And indeed, the Americas have plants that seem to require passage through a large herbivore for optimum germination.

Then humans arrived in the New World, spotted the giant herbivores, thought, "Dinner!" and gobbled them all up. (Well, not necessarily. Some controversy exists about the exact role that humans played in the extinction of these large animals. But the current thinking is that our species did play a significant role — and spent the next few centuries trying to come up with a suitable antacid.) Regardless of how the gomphotheres disappeared, disappear they did, leaving plants without the herbivores needed to facilitate germination of their seeds. In fact, some species of plants in the tropics may have been undergoing a gradual decline in population as a result of an absence of these seed-dispersal agents.

Then along came horses, which aren't native to the Americas. As it turned out, horses also serve as excellent seed dispersers for some American flora, and in some cases, they are the only animals able to facilitate the germination of native plants. But the plants didn't evolve in the presence of horses, which only just got to the New World a few hundred years ago.

Scientists can examine selection for increased running in the laboratory. Of course, it's tough to do experiments with big cats in the laboratory: You need a lot of room for them to run, feeding them is expensive, and if you're not careful, you could end up on the dinner menu. But cats are not the only things that run. Mice run, too. And they're a lot cheaper to keep and a lot less likely to turn on you. And the best part? They're happy to run in those little wheels, so with a minimum of analytical equipment, you can keep track of how far they're running.

Theodore Garland, Jr., and coworkers set up a long-term evolution study looking at the consequences of artificially selecting for mice that like to run. They got a bunch of mice, gave them all a chance to run, and then founded the next generation by selected the ones that just plain liked to run more. They've continued this selective regime for almost 15 years, continuing to selectively favor the mice than most like to run.

The results: Mice lineages selected for increased voluntary running now run everyday several times farther than the original mice, and they do it by running for just as long but several times faster. These fast running mice have undergone other changes as well as a result of this artificial selection:

- ✔ They have larger hearts.
- ✔ Their muscle fibers are different from the original unselected mice.
- ✔ Their hind limbs or more symmetrical.

There was plenty of variation for characters associated with running on which artificial selection could act. It didn't take long to produce evolved mice that like nothing more than to go to the gym for a workout.

Oh — one last thing: The hearts of the mice that ran farther aged better, so when you're done with this book, go take a walk or jog. It'll be good for you.

# Darwin's and Grants' Finches

When Darwin was visiting the Galapagos Islands, he came across a strange assortment of birds which subsequent investigation revealed to be an assortment of different kinds of finches. These finches were so different from the finches found on the mainland that Darwin didn't even recognize them as finches and, while on the Galapagos Islands, didn't pay that much attention to them. It was only when he returned to England and shared his collection with fellow scientists that he realized what he'd missed. But it didn't take him long to realize the significance of the little beasties.

The finches on different islands have different morphological characteristics — things like the size and shape of the beak — that might reasonably be thought to have something to do with feeding. Darwin hypothesized that different selective pressures on the different islands had led the birds to diverge from each other morphologically. Specifically, he surmised that natural selection was at work (refer to Chapter 2).

Darwin hasn't been the only scientist interested in finches. Rosemary and Peter Grant of Princeton University are, too. And what's more, they set out to watch natural selection happen, and they've been doing it successfully for about thirty years.

In one particularly nice example, one of the larger finches which had not previously been present on the island where the Grants work took up residence there and began to compete for food with the smaller finch that was already there.

To grasp the significance of what happened next, you need to know a little bit about the structure of finches' beaks. A finch's beak structure determines what the bird can most efficiently eat: A bird can't crack large seeds with the tiny beak, and it can't pick up tiny seeds with a large beak. Both the larger and smaller finches had a preference for some large, yummy seeds found on the island. And both have bills of sufficient size to crack and eat these seeds.

Prior to the arrival of the larger finch on the island, the small finches had all the big nuts to themselves. When the larger finches arrived, they took over the prime feeding areas as a result of bigger beak size and hogged all the good food. This led to tough times for the smaller birds, especially during periods of low food availability; they weren't getting enough of the big seeds, and they couldn't easily forge on the smaller seeds. Tough times, and a lot of the smaller birds didn't make it, but some of the smaller birds had an easier time of it than others.

Which of the smaller birds were most successful? The ones with slightly smaller beaks. There was variation in beak size, and in the absence of the large seed food resource, the small finches with smaller beaks were able to get more food because they could more efficiently forage on smaller size seeds. As a result of this selective pressure, the big size of the smaller finch decreased. And that's natural selection in action!

# Chapter 6

# Random Evolution and Genetic Drift: Sometimes It's All about Chance

. . . . . . . . . . . . . . . . . . . . . . . . . . . . . . . . . . . . . . . . . . . . . . .

## In This Chapter

▶ Understanding genetic drift: What it is and when it's important

▶ Seeing how genetic drift reduces variation

▶ Navigating the adaptive landscape

. . . . . . . . . . . . . . . . . . . . . . . . . . . . . . . . . . . . . . . . . . . . . . .

harles Darwin (see Chapter 1) had natural selection nailed, even though he didn't have the tools to test his ideas that evolutionary biologists have today. A great deal of modern evolutionary biology has been about confirming and refining Darwin's hypotheses. Since Darwin, one of the most important advances has been the recognition of the role of genetic drift as an evolutionary force. *Genetic drift* refers to the power random events can have in influencing whether genes increase or decrease in future populations.

Here's the take-home message of this chapter: Genetic drift can result in evolution, even in the absence of natural selection. If two critters are equally well suited to their environment, only chance determines which one leaves more descendants. Genetic drift also can work in the presence of natural selection. Even when some individuals are potentially better than others, chance events still occur. If lightning strikes the fastest cheetah, it won't be contributing its genes to the next generation.

## Genetic Drift Defined

A main principle of evolution by natural selection is that the environment favors traits that make individual organisms more fit (better able to get their genes into the next generation). More-fit parents produce offspring who, by virtue of having inherited traits from their parents, are more fit. The most fit of these offspring produce offspring who are even more fit than they, and so on and so forth until . . . well, you get the picture. These changes are not random. More-fit characteristics increase in frequency in subsequent generations at the expense of less fit characteristics.

Here's the rub: Natural selection isn't the only force that determines what genes get into the next generation. It can be doing its thing, stacking the cards in favor of certain characteristics, and then — *bam!* — a tree falls on the genetically favored organism. In the words of the philosopher Dylan, blame it on a simple twist of fate. If that tree happens to fall on a fast cheetah, there'll be just a few more gazelles for the slower cheetahs to eat and they may leave more descendents than they otherwise would have.

When random processes affect the probability of different traits being present in the next generation, you've got what evolutionary biologists call *genetic drift.* And it's one of the two major driving forces changing the frequencies of existing genes through time (natural selection being the other). As such, it can

- ✔ Affect whether the frequencies of different alleles increase or decrease in subsequent generations.
- ✔ Result in the frequencies of all but one allele at a given locus decreasing to the point that they are eliminated. When only a single allele is found at a given locus, that allele is *fixed;* its frequency has gone to a hundred percent and all the other alleles are gone. Until such time as a mutation happens to generate a new allele at this locus, all individuals in the population will be genetically identical at this point in their DNA. No further evolution is possible because, until such a mutation occurs, there is no variation.

The key to understanding genetic drift is to understand what could possibly be random in evolution (quite a bit, actually) and when these random events are evolutionarily significant (sometimes).

Natural selection and genetic drift aren't either/or processes. Think of the processes as happening simultaneously and the circumstances (such as population size or the neutrality of a mutation, as explained in later sections) determining which process holds more sway.

# *Wrapping Your Head Around Randomness*

*Random,* as you no doubt are aware, means out of the blue, without rhyme or reason, hit or miss — in a word, arbitrary. Random events, by their very randomness, exhibit no discernible pattern, even if they look like a pattern; hint at a pattern; or, gosh darn it, *must* form a pattern if you could just figure out what the pattern is. That's precisely the part about randomness that trips us humans up: We're always looking for a pattern, and if we don't see one, we expect it to emerge eventually.

Well, randomness doesn't work that way. Random is random, and that means:

- ✔ The absence of a pattern or the unpredictability of an outcome
- ✔ No correlation between the outcome of one event and the outcome of another

A coin toss is an example of a random event. You toss a coin in the air (assume it's a fair coin), and it's going to come down either heads or tails. So say you toss this coin, and it lands heads. What's the chance that it'll land heads the next time you toss it? The right answer is 50 percent. But a fair number of people think (incorrectly) that it is more likely to land tails the second time because it landed heads the first time. This sort of thinking is the same kind that makes gambling so dangerous.

Imagine yourself, flush with coins, sitting at a slot machine and feeling lucky. You drop in your first coin, pull the lever, and . . . nothing happens. You put in the next coin. Again, nothing happens. And again. And again. Why do you keep playing? Probably at some level, you think that all these losses are leading up to a big win, even though you may be fully aware that a slot machine generates numbers or patterns randomly.

The important thing about the coin toss (or the slot machines) is that, as a random process, the previous outcomes don't have any bearing on the upcoming results. Just because a coin lands heads one time doesn't mean that it's less — or more — likely to land heads the next time.

Random is just plain slippery to think about so following are a couple of examples of how it can be important in biology.

## At the level of the individual

Any environmental factor that affects an individual's ability to reproduce regardless of its genes can cause genetic drift. Consider a lightening strike: There's probably no genetic component that determines whether one deer or another is hit by lightning. But when a deer *is* hit by lightning, it won't be reproducing. Because that deer's genes won't be represented in the next generation, this random event changes the relative representation of genes in the later generation.

Essentially, these are cases of being in the wrong (or right) place at the wrong (or right) time. Examples would be a lumberjack who fells a tree on a Nobel laureate who's walking through the forest pondering his acceptance speech or the bug that crashes into your windshield. If you're not around to reproduce, or if you don't reproduce (because you prefer a neat house and travel to children, for example), natural selection can't be the driving force behind the increases or decreases in gene frequencies.

Random variations also occur in the number of offspring that different individuals have. For the minute, ignore the fact that there might be a genetic component to wanting fewer children because you prefer not stepping on those little plastic farm animals with the sharp ears. Random variation still occurs in the number of offspring that different people produce. A hundred couples all trying to have three children won't each end up with three children. On average, they'll end up with more children than couples who prefer not to have children (a matter of choice, not random factors), but they won't all end up having exactly three. That's where the random factors come in.

## At the level of the gametes

As a diploid individual, you have two copies of DNA: one that you got from your father and one that you got from your mother (refer to Chapter 3). Your children will also have two copies of DNA: one copy that they got from you and one copy that they got from the other parent. That means that each of your offspring got only half of your genes and half of their other parent's genes. You're diploid, your partner is diploid, and the only way to make sure your offspring are diploid is for your gametes (the egg and sperm) to be haploid — that is, to contain only half of your DNA.

Which half? This is where genetic randomness kicks in. At some locations of your genome, both copies are the same. Or in science-speak, both alleles at that locus are identical. In this case each of your gametes (be that sperm or egg) will have the same allele at that locus; there's only one to choose from. But other times, you have two different alleles at that locus because the one you got from your mother and the one you got from your father weren't the same. Any given gamete you produce will get one or the other.

You've been successful: You've survived (you looked both ways before you crossed the street), you found food (maybe that was why you crossed the street), and now you're about to pass your genes on to the next generation, but it's a matter of chance which of the two different alleles you carry will make it into any particular gamete. On average, each allele will end up in half of your gametes. While you produce a huge number of gametes and the two alleles are equally represented, only a paltry number of offspring are likely to result from them.

Women have about 400,000 eggs; men produce even more sperm. But women don't have 400,000 offspring. For that reason, which particular genes your offspring have are up for grabs. In other words, the production of any one child involves an element of chance, and the kids you end up with are the result of random pairings, as it were.

# Situations in Which Drift Is Important

Genetic drift is an important evolutionary force — some of the time. When? When populations are small, for example, or with the subset of genetic differences that don't result in fitness differences.

## When a population is small

Genetic drift plays a bigger role in evolutionary change when a population is small. In larger populations, the forces of genetic drift are muted. In a population of 1 million, one extra fast cheetah more or less isn't going to make much difference in the grand scheme of gene frequencies. In a population of 20, however, that single cheetah *does* make a difference — at least in terms of her genes — if she does (or doesn't) reproduce.

Consider the coin toss again. If you throw up a bunch of coins — say, 400 — the outcome would be lots of heads and lots of tails — about half of each, in fact. Although you'd be willing to accept a few more or less than half either way, you know that only a very small chance exists that all 400 coins will end up falling on the same side. Now imagine that you're tossing a small handful of coins — say, four. The possible outcomes are fairly limited: You could get two of each, three of one and one of the other, or four of either heads or tails. Although you're not likely to see the coins come up four of a kind, that result is still within the realm of possibility. In fact, if you compute the actual probability, you realize that, on average, one of every eight times you throw four coins in the air, they'll end up on the same side.

The point? That even though a handful of coins will on average land half heads and half tails, that's just the average result. Individual tosses will vary randomly, and the smaller the handful, the more likely the chance that the results will be much different from 50-50.

Now think about allele frequencies rather than head-tail frequencies. A population may have two alleles in equal proportion, 50 percent each, at one point in time. And then, at some later point in time, the frequencies are drastically different. The smaller the population, the more likely that this change is simply the result of random factors affecting which individuals did and didn't leave more descendents.

# When genetically different individuals have the same fitness

The degree to which a new mutation is neutral affects how important a role genetic drift plays in the evolutionary process. As Chapter 4 explains, mutations, which add genetic diversity to populations and come in three categories (advantageous, deleterious, and neutral), are always occurring.

For mutations that are advantageous or deleterious, and when population size is large, natural selection is the primary driving force that determines whether the frequency of particular alleles increase or decrease in subsequent generations. If mutations are deleterious, natural selection removes them; if they're advantageous, natural selection favors them, and they increase in frequency.

But if the particular new mutant is *neutral,* neither advantageous nor disadvantageous compared to the original, natural selection can't be responsible for changes in its frequency through time. In this case, genetic drift is the driving force.

# Drift or selection? When it's hard to tell

Genetic drift is always occurring, even when an allele is advantageous or disadvantageous. Randomness is still at work — lightning is still hitting the occasional deer, for example — but these random events don't make much difference to the outcome. The slow deer will get eaten by wolves, and the fast deer will pass their genes on to the next generation. If a lightning bolt happens to hit one of the fast deer, the outcome won't change. But sometimes, it's not so clear cut when natural selection is the main force and when genetic drift is.

## The effect of population size

In a small population, the weakly advantageous allele could be eliminated by random genetic drift before it could be fixed by natural selection, or a weakly deleterious allele could increase to fixation. For example, a cheetah with a mutation that made it just a tiny bit faster might be expected to catch a few more antelope and have a slightly better chance at leaving more descendents than its neighbor. This new mutant would tend to increase in frequency in the population through time, but it would increase very slowly because the difference between the mutant gene and the other cheetahs is miniscule.

It might also increase or decrease in the population purely as a result of chance events. It's possible that even though it's selectively advantageous (you get to catch a couple more antelope, just not very many more), it might be eliminated by chance events before it could sweep through the population.

Whether a particular neutral allele increases to fixation is a function of population size. In small populations, random factors can be of greater importance. But — and this is a big *but* — the probability that there will be a neutral mutation that will increase to fixation is *independent* of population size. The key is that larger populations have more individuals and, as a result, more errors in DNA replication (errors by the mutations). So although it's true that any particular mutation has a much lower chance of increasing to fixation in a large population, that population has proportionally more mutations. Mathematically, everything pretty much cancels out. The effect of large population size on slowing fixation as a result of genetic drift is balanced out by the fact that a large population also has more mutations.

Here's why this little tidbit is important: Knowing that the overall rate of accumulation of neutral mutations doesn't have anything to do with population size lets scientists estimate the time since two lineages diverged. Neutral mutations in particular genes sometimes accumulate at a relatively constant rate. Therefore, the differences between two lineages can be used to determine the time since their divergence — a subject covered in more detail in Chapter 15.

### The strength of selection

Natural selection doesn't operate at the same strength at all times. Imagine a gene that allows an organism to survive extremely dry conditions. Natural selection will result in an increase in the frequency of this gene in environments with very dry conditions. Now imagine that those conditions occur only once every ten generations. During the other nine generations, changes in the frequency of that particular gene from generation to generation will be due entirely to drift.

If a new allele is neutral, random forces determine its future frequency. It may disappear, or it may increase in frequency. Consider the gene that determines which of your thumbs is on top when you interlace your fingers. As it turns out, you always put the same thumb on top. Try interlacing your fingers so that the other thumb is on top. Feels weird, doesn't it? Yep, a locus with two alleles actually determines your thumb preference.

Assuming that both alleles have always been identical with respect to fitness (scientists can't be certain, but neither can they think of the slightest reason why they wouldn't be), both alleles exist in the human population because of random events — an initial mutant increasing in frequency as a result of chance in the absence of any effects of natural selection. The relative frequency of the two genes is still subject to the process of genetic drift, but human populations are so large that both alleles will persist in the population for the foreseeable future.

# Genetic Drift in Action: When Big Populations Get Little

Understanding the importance of population size is fundamental to understanding genetic drift (see the earlier section "Situations in which drift is important" for details). But population sizes don't remain constant; they fluctuate over time, sometimes growing larger and sometimes smaller. To understand how genetic drift is operating in nature, it helps to understand the factors that result in a small population. Here are a couple:

- **Fluctuations in population size:** The mere fact that a population is large today doesn't mean that it wasn't small at some other time or that it won't decrease someday. The populations of many organisms that need water may drop to low levels during a drought, for example. When a population is or becomes exceedingly small, the phenomenon is called a *population bottleneck* (discussed later in this chapter).

- **Founder effects:** Founder effects refers to those chance events that occur when a population is founded in a new location by a small number of individuals from a population somewhere else.

As a population shrinks (which can happen for a variety of reasons), it begins to feel the effects of genetic drift more acutely, mainly because while it's growing smaller, it's losing diversity. This loss is a bad thing, because diversity can protect a species or a group of individuals from the vagaries of fate. This discussion is important for a couple of reasons, one of which may appeal to folks in general (tree-huggers and animal-lovers in particular), and the other may appeal primarily to scientists (but is cool anyway):

- **It informs conservation efforts.** Numbers aren't enough when it comes to saving animals from extinction. Diversity in the endangered population is key, and genetic drift undermines diversity.

- **It lets scientists figure out when one group diverged from another.** I know — this reason lacks the pizzazz of saving cute little furry creatures, but it's still important.

The following sections go into more details.

## Population bottlenecks

*Population bottleneck* simply refers to a period when the population of a particular organism grows smaller. If, as the population decreases, it also loses genetic variation through genetic drift, it doesn't get that variation back right away when (or if) it gets bigger again.

For an example of a population bottleneck, consider the plight of the northern elephant seal. This seal was hunted almost to extinction in the late 1800s. On a couple of occasions, folks thought that the species had actually gone extinct, but a few seals occasionally turned up on an island off Baja California. Most of these seals were killed as well (sometimes by biologists collecting them for museum specimens!). Obviously, things weren't going well for the northern elephant seal; at the species' low point, the population was estimated to be between 20 and 100 individuals.

In the early 1900s, the Mexican government made the seals' last breeding ground a protected area, and the species began to rebound. Today, the species numbers over 100,000, and elephant seals appear to be out of immediate danger of extinction. This example is a nice success story of species conservation in a world that has too many unhappy endings.

Before you get out the party hats, however, remember that genetic variation is necessary for evolution by natural selection. If the species we humans conserve are going to respond to future environmental changes, those species require genetic variability. The world may have 30,000 elephant seals now, but those 30,000 seals are descended from the few that survived hunting. That means they're very closely related to one other and, as a result, are genetically very similar, so they are still at risk from events such as diseases. In a genetically uniform population, a disease that can attack one of the individuals is likely to attack them all.

The southern elephant seal was also extensively hunted, but its population was not driven as low at that of the northern elephant seal. As a result, the southern species has more genetic diversity than the northern species.

The genetic diversity of a species is a fundamental part of what is unique about that species. Yet the forces of genetic drift can eliminate genetic diversity in small populations, and small populations are often characteristic of species that humans are trying to prevent from going extinct. Therefore, counteracting the effects of genetic drift — and not inadvertently creating more population bottlenecks by using too few individuals for breeding in captive breeding programs — is important in species-conservation efforts. Genetic diversity (heritable variation) is what allows species to respond to selective events (such as novel diseases or environmental change). Conservationists want not just to conserve a few individuals of a species but also to conserve the genetic variation that will allow the species to survive events it may encounter in the future.

# Founder effects

Periodically, a few individuals decide to strike out on their own and set up a community away from kith and kin. The founders of this new community — be they rats, bats, bears, or humans — determine the genetic makeup of the new population. *Founder effects* refers to their influence in this regard. Understanding founder effects helps scientists understand how the genetic makeup of original members of the new community affects the gene frequencies in subsequent generations.

Every time a few individuals separate themselves from their original population and move to a novel environment, you can see how founder effects can be important in determining allele frequencies. Quite a bit of variation in human blood type occurs across different geographical regions, for example. Some areas have a preponderance of one blood type; other areas have a preponderance of another type. Founder effects, spurred by human migration patterns that often involve small groups founding new populations, can explain these differences, as the following sections explain.

## Crossing the Bering Strait

Ten thousand years ago, a small group of humans crossed the Bering Strait and colonized North America (and subsequently South America). Genetic drift probably is the cause of the very low frequency of the type B blood allele in Native Americans (4 percent), even though it is much more frequent in people from Asia — the presumed source of the Native American colonists.

## Amish immigration to America

The Amish population was founded by a small group of Europeans in the early 16th century. In the 18th century, the first Amish migrated to the Pennsylvania colony in the New World.

The Amish, who remain almost entirely reproductively isolated from the rest of America, have a distribution of blood types different from those in the rest of the United States or in Europe. Additionally one of the original couples carried the allele for polydactily (extra fingers or toes) which has increased in frequency in the small Amish population by chance; they just happen to have more than the average number of descendants.

## Colonization of planet Beta

Okay, this example isn't for real. But if you've seen the 1950s-era sci-fi flick *When Worlds Collide,* you can imagine founder effects in action. In the film, Earth is doomed by the approach of a runaway planet. Humanity is saved when a group of scientists and engineers (including the requisite cynic-turned-true-believing hero, the brilliant-and-self-sacrificing old scientist, his beautiful-but-prone-to-melodramatic-emotional-displays daughter, and the steady-but-dull-as-dishwater-and-soon-to-be-spurned boyfriend) gets together to build a spaceship large enough to carry a few humans, a few head of livestock, a few plants, and one adorable dog to another planet to begin life again. Think of the genes this bunch has to work with! Their blood type — given that they left the riffraff on Earth to perish — definitely would be blue.

# The Shifting-Balance Hypothesis: It's What's Wright

Sewell Wright (1889–1988) was a key founder of the science of population genetics, and his shifting-balance hypothesis is one of the coolest hypotheses about how evolution might work in certain circumstances. This hypothesis says natural selection and genetic drift can work together to allow populations to reach higher fitness.

To help visualize how this might work, you need to get cozy with an especially slippery concept called the *adaptive landscape* (or *fitness landscape*). It's slippery even for professional evolutionary biologists, but of course that's part of the fun!

## The adaptive landscape: A 3-D fitness map

In any giving species, there are a huge number of different possible genotypes, and we can't possibly know the fitness of each different genotype — we just know that each genotype has a fitness. In order to think about all that information, scientists use the *adaptive landscape*. The adaptive landscape is a three-dimensional imaginary structure that gives us a way of talking and thinking about how differences in genotype affect fitness. Figure 6-1 is an example of a fitness landscape. The letter A represents a mutation that's beneficial (moving the species up the hill to greater fitness); the letter B represents a mutation that decreases a species' fitness.

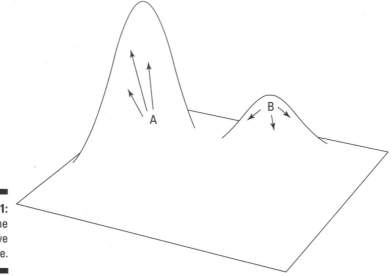

Figure 6-1:
The
adaptive
landscape.

This landscape is like a 3-D relief map that you can set down on your coffee table. A relief map indicates, through the height of the bumps, the altitude of any region — such a map of the U.S. would show the Rockies and the Sierra Nevadas, the Midwestern plains, the small mountains of Appalachia, and so on. The adaptive landscape has the same bumps, but here, the bumps don't show altitude; instead, they correspond to increased fitness.

Here are some things to know about the adaptive landscape:

- **Each location on the adaptive landscape can be though of as a genotype.** Neighboring points are very similar genetically, distant point less so.

- **The adaptive landscape deals with one species at a time and in one set of conditions.** Imagine the adaptive landscape for chickens on an island in the Pacific, for example. For this particular population, some genotypes confer greater fitness and other genotypes confer lower fitness.

- **The top of each bump on the adaptive landscape represents a genotype more fit than those just a little different from it.** The different heights of these bumps indicate relative fitness. Higher bumps equal higher fitness levels; lower bumps equal lower fitness levels.

Any particular adaptive landscape represents the fitness of a single species in a particular set of conditions. Move to a different island and the fitness landscape might change. Genotypes that were great on the first island might be hopeless on the next, and therefore, the peaks and valleys would be in different places.

It's impossible to measure the fitness of every genotype of the species being studied. But you can think about how mutations change fitness. A mutation that switches an amino acid from one kind to another, for example, nudges you a little bit on the adaptive landscape. Maybe the protein works better (you move up the hill); maybe it doesn't work as well (you move down the hill); or maybe it works exactly the same way (you move sideways on the hill, meaning that you're in a different place — you have a different genotype — but you have the same fitness. Your genome could change in millions of ways as a result of mutation; it's a big genome. Each of those changes moves you around somewhere in the fitness landscape.

## Being the best you can be — on your own peak

You expect that natural selection will favor the genotypes that make an organism more fit and select against those that make it less fit. For example, plants with better roots get more water and make more seeds, and natural selection drives the population to have better roots from one generation to the next. But maybe there are different ways to make a good root (or anything else in the biological world).

The adaptive landscape is a way of thinking about the fact that several peaks of high fitness may exist, each peak corresponding to a different genotype. Natural selection will drive a population to the top of whatever peak it's already on, whether or not that peak is the highest one in the adaptive landscape.

What natural selection *won't* do is move the population to an even higher peak if that move means crossing a valley of lower fitness. Why? Because natural selection won't backtrack by favoring individuals with lower reproductive success. Individuals with lower fitness arise all the time (remember that most mutations are bad), but they don't increase in frequency as a result of selection.

## You just can't get there from here

Sewell Wright's key insight (one of his many key insights — he's one of the hot shots of post-Darwinian evolutionary biology) was suggesting that genetic drift is the mechanism that lets a population move from one fitness peak (one really great genotype) to another because, unlike natural selection, genetic drift lets populations wander around low-fitness parts of the adaptive space. In his hypothesis, a small population could descend (genetically speaking) into an adaptive valley (an area of low fitness) and then climb a different, higher, adaptive peak as a result of natural selection. Essentially, a small population has a chance of hitting the genetic jackpot and ending up on a high adaptive peak.

Genetic drift is like a lottery; no guarantee exists that genetic drift is going to make it possible for a species to wander over to another peak. Wright's argument is just that it can do so *some of the time*. The low part of the fitness landscape means that those are genotypes that have a lower chance of leaving descendants. But they might get lucky once in a while.

When a small population does hit the genetic jackpot and gets to the slope of a higher peak, natural selection will kick in again and the population will climb up the new higher peak.

With a really fit combination of alleles, the population can spread through the *actual* physical landscape. Because the individuals have higher fitness (they're at the top of a really high fitness peak), they increase in frequency and can migrate in the physical landscape, spreading their selectively advantageous genetic combination far and wide. In other words, one small population hits the fitness jackpot; then these individuals sow their oats in other populations.

# Chapter 7

# Quantitative Genetics: When Many Genes Act at Once

## In This Chapter

▶ Teasing apart phenotypic variation into genetic and environmental components

▶ Making sense of broad- and narrow-sense heritability

▶ Measuring the response to selection on quantitative traits

Many of the traits scientists are interested in aren't the result of a single gene, but of multiple genes. As discussed in previous chapters, some traits are controlled by one gene (or by a known few genes). In contrast, *quantitative* (or *multigenetic*) *traits* are determined by multiple (more than two) genes.

When multiple genes and environmental factors, which can have a big impact on how traits are expressed, both play significant roles in the phenotype, how much of the phenotype depends on the genes (and therefore is heritable), and how much depends on the environment (and is not heritable)?

If you expect that the next question is "Why in the world would anyone care?" remember that in evolution, only heritable variation is important. Many of the traits scientists, agriculturalists, and medical professionals care about turn out to be quantitative traits. The field of quantitative genetics deals with determining what proportion of the variation is due to genetic factors when multiple genes and the environment play a role.

Geneticists in this field use all sorts of mathematical formulas and statistical techniques to figure out all this information, none of which you need to know. So this chapter skips the math and focuses on the key principles instead. Here, I paint with a broad brush but give you the foundation you need to understand this topic — just in case you run across it in the science section of the newspaper.

# Why Quantitative Genetics Is Important

Quantitative genetics is relevant to evolution because it has to do with heredity and understanding how heredity works. Darwin didn't understand how heredity worked; he just knew that it did — sort of. Back in those days, folks understood that the heritability of some traits was more predictable than others. For some things, such as hemophilia, researchers could trace the ailment back to a single factor that they could map onto a family tree. (Today, scientists know that this factor is a particular DNA sequence.)

Other traits, such as how much milk a cow made, weren't so clear cut. True, researchers in Darwin's day knew that things like how much milk a cow made had a heritable component, but figuring out what that heritable component was wasn't as simple as determining the heritable component of eye color or a disease like hemophilia.

Fast-forward several generations. Scientists now can explain some of the things Darwin and his contemporaries could only wonder about. We now know that a character is often the result of several genes rather than a single gene.

## Interacting genes

When a whole bunch of genes affect a particular trait, these genes can interact in different ways. The effect of the different genes can be additive, though some genes might be more important than others in determining the final phenotype. Imagine a lot of genes for milk production where, at each locus (location on the DNA strand), some alleles result in more milk and some result in less. The heritable part of milk production will be determined by the sum of the effects at all the involved loci.

Different loci can also interact in non-additive ways. In some cases, the non-additive interaction of the different genes is important. Imagine that a particular locus may influence milk production, for example, but only when a specific allele occurs at some particular other locus. Without that allele at the second locus, the first locus doesn't have anything to do with how much milk a cow makes. This process is called *epistasis*.

*Epistatic* interactions are those in which the fitness of a particular allele at one locus depends on alleles at another locus.

Imagine epistatic interactions that affect the A locus, in which the relative fitness of the aa and AA individuals depends on the alleles at a second locus, which I'll call B. If that second locus has one combination of alleles, AA individuals will have higher fitness than aa individuals. But if the B locus has a different combination of alleles, aa individuals will have higher fitness than AA individuals.

I made up the milk example for illustrative purposes, but one actual case of epistasis that you might be familiar with concerns those colored squash that you see around Halloween and Thanksgiving. In one genetic system, the colors white, green, and yellow are controlled by two genes, each of which have two alleles (of the dominant/recessive type; refer to Chapter 3).

When at least one of the genes at the white locus is the dominant form, the squash is white and the other locus has no effect on color. But when both alleles at the white locus are of the recessive form, then the color of the squash (green or yellow) is determined by the alleles at the green/yellow locus. You can see that it starts to get complicated even with only two loci involved — there can be many more.

These examples have consequences for the way selection operates. In both cases, the extent to which selection increases the frequency of the alleles at the one locus depends on the frequency of the alleles at other locus.

## Multigenetic traits in medicine and agriculture

Being able to figure out the correlation between heritable genes and phenotype when the genetics is more complicated and environment plays a role has several benefits. Multigenetic traits are important to the fields of medicine and agriculture, for example.

### In agriculture

Most of the characteristics that farmers breed for are quantitative traits, such as the amount of milk a cow produces or how many ears of corn a corn plant produces. To help farmers breed better cows and better corn plants, researchers need to understand what's involved when they start thinking about traits controlled by many genes.

### In medicine

Understanding genetic interactions can help scientists sort out the genetics of human disease. For many diseases, the heritable component is not absolute. People say that a disease or condition tends to "run in families" or that you're more likely to have the condition if your recent ancestors had it. Both expressions indicate that you may be predisposed to a particular condition that has a genetic component, but you won't definitely get it. Maybe the ailment appears only under certain environment conditions or only when certain genes interact in a particular way.

# *Understanding Quantitative Traits*

Although all quantitative traits are multigenic traits, not all multigenic traits are quantitative traits. Strictly speaking, a multigenic trait is any trait controlled by more than one gene, yet when only a couple of loci are involved, researchers can sort out all the different combinations of outcomes and understand the phenotypes, as in the squash color example in the earlier section. As the number of genes gets larger, doing that isn't so easy, so scientists have to use the mathematics of quantitative genetics.

## *Continuous and non-continuous traits*

Quantitative traits fall into two camps: those that vary continuously and those that don't:

- **Continuous quantitative characters:** Quantitative traits often are continuously variable across some range. An example is adult height. Although you may consider yourself to be either tall or short, heights vary smoothly from the tallest to shortest.

- **Non-continuous quantitative characters:** Although many multigenetic traits are continuous, some aren't — at least, not in the same way that continuous characters are. Take bristles on fruit flies, for example. Fruit flies have a discrete number of bristles that varies from some bristles to more bristles, but they don't have half a bristle. Your weight can vary by any fraction of a pound (or gram or stone), but a bristle or a pound is either "on" or "off." Some diseases or conditions (those that have two states: sick or not sick) are also non-continuous characters.

The environment has a big impact on many quantitative traits. Taller parents, for example, tend to have taller children, but environmental factors, such as nutrition, also affect adult height.

## *Crossing a threshold*

For a non-continuous quantitative trait to be expressed, some threshold combination — of genetic interactions alone or of genetic interactions combined with environmental factors — has to occur.

Take mental illness, which scientists are still trying to understand. Researchers hypothesize that disorders such as schizophrenia are controlled by multiple loci and that the condition manifests in people who have a certain number of specific alleles at specific loci (and perhaps in certain environments), but whether the condition manifests itself depends on more than the genetic component.

## The case of hemophilia

In the case of hemophilia, several different loci code for different clotting factors, and different forms of hemophilia result from alleles (that code for bad clotting factors) at several of these loci. But hemophilia isn't a multigenetic character, even though several genes can be responsible for it. Why? In each case of hemophilia, a single locus is responsible for the trait. Which one you just happen to have been unlucky enough to inherit determines the type of hemophilia you have.

Here are a couple of other interesting tidbits about hemophilia:

✔ It's an X-linked recessive character, which means that two copies are necessary for a woman to have hemophilia, but only one copy is necessary for a man to have it.

✔ As a result of modern medicine, hemophilia has gone from a trait associated with very low fitness to one with greatly reduced fitness consequences. Having hemophilia is never good, but it's not as bad as it used to be because today medications can replace the lost clotting ability. This situation is yet another example of how the fitness consequences of a particular gene are a function of the environment.

Schizophrenia, for example, definitely has a genetic component. If one of your parents has schizophrenia, you're at higher risk of developing schizophrenia than someone in the general population; if both your parents are schizophrenic, your changes are even greater. This situation makes sense, given that you share some, but not all, of your parents' genes. But here's a particularly interesting point that reveals that genetics isn't the only factor: If your *identical* twin is schizophrenic, you have about a 50 percent change of developing the disease yourself, even though you share 100 percent of your twin's genes.

Although the genetic component is significant, something beyond genetics is going on. Maybe environmental factors are the key. In the case of relatives other than identical twins, perhaps different combinations of genes and their interactions are important as well. These questions are what schizophrenia researchers are trying to sort out.

# QTL mapping: Identifying what genes matter

When you're dealing with multigenetic traits, figuring out what genes are responsible for a particular trait is harder but not impossible. The strategy for identifying the relevant genes in the genome of a particular organism involves *quantitative trait loci mapping* (*QTL mapping*, for short). With QTL mapping, you cross lots of individuals and keep track of how particular genetic markers in the offspring correlate with the character you're interested in. These markers aren't themselves the genes responsible for the quantitative trait, but because they're correlated with the phenotype, they can be used to identify the location on the chromosomes where the genes reside.

A *marker gene* is any gene that's easy to recognize because its location on the genome is known. Imagine, for example, that people with green eyes are on average taller. We know that tallness is a function of many genes, but now we know that one of these genes is near the gene that controls green eye color. The two genes are in close proximity. They are closely *linked* — they tend to be inherited together. (Go to Chapter 5 or more details on linkage.)

Here's how QTL mapping works: Scientists know where the marker gene is, and now they know that some gene near it is responsible in part for a multi-genetic trait. The mathematics of how this works is beyond the scope of this book, but what's important is that it does work. (And it works even better for species for which we have a lot of gene sequences where we can go in and see exactly which genes are near the marker loci!)

By using the markers as signposts, scientists can

1. Figure out approximately how many genes underlie a given multigenetic trait

2. Tease apart the details of gene interactions

3. Understand the relative importance of different genes to see which genes have stronger effects than others

4. Voila! Actually have a good idea where to look in the genome for the relevant genes.

# Analyzing the Heritability of Quantitative Traits

As stated earlier in this chapter, quantitative traits are traits that result from complex interactions between multiple genes and that may be influenced by environmental factors. To understand how these traits evolve, evolutionary biologists analyze the heritability of quantitative traits.

As you can imagine, the first task is to determine what proportion of the trait (or phenotype) is due to genetic factors (the heritable bits) and what portion is due to the environment (the non-heritable bits). As tough as that job is, the analysis is made even more complicated by the fact that any given gene, or allele, may impact the resulting phenotype in an additive or non-additive way.

# Additive or non-additive?

The concept of additivity can be a little slippery. To understand it better, think about a single gene with two alleles: for example, wrinkled or smooth peas. In this case, the smooth form (A) is dominant over the wrinkled form (a). An individual with a single allele for the smooth form (Aa) would express the smooth phenotype. Having two copies of the smooth form (AA) produces the same smooth phenotype. In the first case (Aa), the single recessive allele is non-additive, as is the second dominant allele in the second case (AA). Neither allele affects the final expression of the trait. (For details on dominant and recessive genes, see Chapter 3.)

This situation, wherein the combinations of different alleles can result in some alleles not having an effect on the phenotype, can happen across different loci. A gene may have the potential to influence a particular phenotype, but it may not influence the phenotype in an additive way. In other words, when the conditions are right (a certain combination of alleles across loci or a particular interaction with other alleles), the gene may influence the phenotype. At other times, this same gene may not affect the phenotype, because the necessary combination or interaction didn't occur.

A hot subtopic of quantitative genetics is sorting through the non-additive nature of multiple genes to figure out whether some of them are especially important (or especially important some of the time). Given that some ailments certainly have a genetic component that's controlled by multiple genes, it would be nice if researchers could identify genes that have large effects and then figure out what they do. This is crucially important for understanding the genetics of some human diseases.

# Determining phenotypic variation

Evolution requires that variation exist and that this variation be heritable. The upshot of environmental effects and of the non-additive genetic variation is to *decrease* heritability and, as a result, decrease the power of selection to transform populations. To determine the strength of selection, scientists separate the variation in a population into the differences due to genetics (both additive and non-additive) and the differences due to environment.

Variation is a property of *groups* of individuals or populations, not individuals alone, no matter how fickle, unpredictable, or changeable those individuals are. All the analysis performed to determine variation relates to groups of individuals.

To understand all the components of an analysis of the heritability of quantitative traits, consider a simple hypothetical example: height. (Why height? No particular reason. You could use any quantitative trait.)

The phenotype — in this example, whether you're tall or short — is a function of the genotype and the environment. Height has a heritable component: Tall parents tend to have taller offspring. It also has an environmental component: Absent a proper diet, you won't get very tall.

You can further partition the genetic component of the phenotype into additive and non-additive parts. How strong selection is for this trait is determined by the additive component of genetic variation.

Here's the math: The phenotypic variation within a population is the sum of additive genetic variation plus the non-additive genetic variation plus the environmental variation. If you like formulas, here's what this one looks like:

phenotypic variation = additive genetic variation + non-additive genetic variation + environmental variation

## Broad- and narrow-sense heritability

Environmental variation isn't heritable. Imagine two people who have a similar genetic makeup, one of whom is taller due only to a better diet. Because the variation between these two individuals isn't due to genetic factors, the taller person won't have taller offspring. But variation that is a function of genetics and not of the environment is heritable.

For the purposes of understanding natural selection, it's helpful to think of heritability as being either the broad-sense or the narrow-sense type:

- **Broad-sense heritability:** The total of all of the genetic factors, be they additive or non-additive
- **Narrow-sense heritability:** The subset of the genetic component that is additive.

Heritability is measured as a number from 0 to 1, indicating the degree of correlation between the parental phenotype and the offspring phenotype:

- If the offspring phenotype is predicted by the phenotype of the parents, heritability is 1.
- If the offspring phenotype is *not* predicted by the phenotype of the parents, heritability is 0.

In the height example, in which the difference between the height of two people was due simply to diet, the phenotype of the offspring of the tall parent would not be any different from the phenotype of the offspring of the short parent, and heritability would be close to zero.

Think back to the example of the smooth and wrinkly peas. Imagine two pea plants, both of which are heterozygous for the smooth character — that is, each plant has both a dominant smooth allele and a recessive wrinkled allele. Because of the dominant interaction between these two genes, all the peas are smooth. Now suppose that you cross these two pea plants. On average, one quarter of the offspring will have wrinkly seeds. In this very simple case, you can see that although the phenotype of the offspring was a direct result of the genes they inherited from their parents (broad-sense heritability), the phenotype of the parents was an inexact predictor of the phenotype of the offspring. Heritability in the narrow sense was less than 1.

In evolution, narrow-sense heritability is the important form of heritability. Imagine that for some reason, being a pea plant with smooth seeds is advantageous. Natural selection will result in the pea plants with smooth seeds being the ones to leave more descendents, and as a result, the next generation will have fewer plants with wrinkly seeds — but not as many fewer as you would expect based just on the relative selective advantage of having smooth seeds. Why? Some of the plants with smooth seeds will have wrinkly offspring. That phenomenon is the non-additive part of the genetic variation.

# Measuring the Strength of Selection

Evolution by natural selection relies on heritable variation and the strength of selection. For that reason, examining those two different factors is important. Scientists can easily measure how phenotypes vary in a population. What they don't know as easily is how much of this variation is heritable. But they can find out.

Figure 7-1 shows the variation of a particular quantitative continuous phenotypic trait in a population — in this case, height. The measured heights fall along the x-axis. Individuals on the left are shorter than individuals on the right. The y-axis shows the frequency in the population of different phenotypes. In this particular example, most of the individuals cluster in the middle range; some of them are very short, and some of them are very tall, but most of them are in between.

To find out how heritable height is in this population, you can selectively breed the tallest (or the shortest) individuals and then examine the frequency distribution of their offspring's height. Figure 7-2 is the same frequency distribution as Figure 7-1, highlighting the subset of the original population you plan to use to create the next generation in your experimental population. In this case, assume that you picked the tall ones.

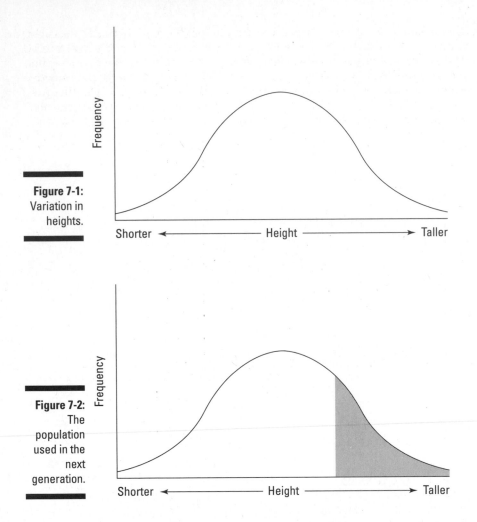

**Figure 7-1:**
Variation in
heights.

**Figure 7-2:**
The
population
used in the
next
generation.

So what will the offspring population look like? Specifically, what will be the frequency distribution of phenotypes of this quantitative trait? Figure 7-3 and Figure 7-4 show two of many possibilities. Each figure shows two frequency distributions. The first is the frequency distribution of the phenotypes of the original population, and the second is the distribution of the offspring population resulting from breeding only the tallest individuals. In both cases, the average height of the population has been shifted to the right. In both cases, on average the offspring population is taller. But in Figure 7-4, the population has shifted much farther to the right, meaning that the offspring population is significantly taller.

A result such as the one in Figure 7-4 tells you that the trait (in this case, height) is much more heritable in that experiment than it was for the case in Figure 7-3. The difference between the two figures can't be due to the strength of selection, because that was the same in both cases. (Remember, you were the selective agent because you picked which ones would have offspring.) Given that the strength of selection was the same, the difference in the response to selection was purely a function of the difference in heritability between these two populations.

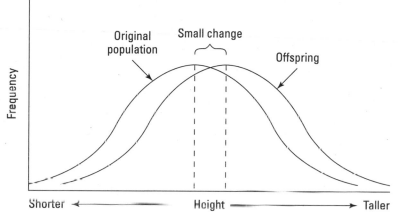

**Figure 7-3:** Phenotypes of offspring from the original population.

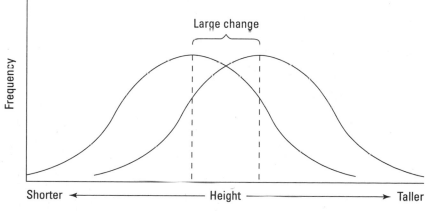

**Figure 7-4:** Height distribution of the offspring of the tallest people.

If you increase the strength of selection, you would expect an even greater increase in the change in average height in the offspring population, which is exactly what Figure 7-5 illustrates. Here, the new population was founded with a much smaller subset of the original population, comprising only the tallest of the tall (compare the shaded area here with the shaded area in Figure 7-2). The result is a greater shift toward an even taller offspring population.

All of this — QTL mapping, continuous/non-continuous and additive/non-additive traits, broad- and narrow-sense heritability — is pretty academic, and you probably need to be a scientist (or a very devoted reader) to grasp the fine and not-so-fine points of the topic. But anyone can appreciate the advantages that the study of quantitative genetics can bring to fields that touch us all. The closer we get to figuring out where all the genes are on the genome, how they interact, and what they actually do, the closer we get to treating diseases that confound us now.

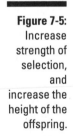

**Figure 7-5:** Increase strength of selection, and increase the height of the offspring.

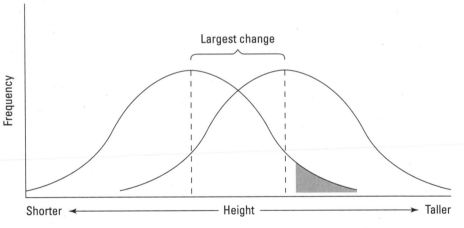

# Chapter 8

# Species and Speciation

● ● ● ● ● ● ● ● ● ● ● ● ● ● ● ● ● ● ● ● ● ● ● ● ● ● ● ● ● ● ● ● ● ● ● ● ● ● ● ● ● ● ●

## In This Chapter

▶ Figuring out what a species is

▶ Surveying the various mechanisms that can limit gene flow

▶ Mastering the mechanisms of reproductive isolation

● ● ● ● ● ● ● ● ● ● ● ● ● ● ● ● ● ● ● ● ● ● ● ● ● ● ● ● ● ● ● ● ● ● ● ● ● ● ● ● ● ● ●

*E*volution is nothing more than changes in gene frequencies in a group of organisms through time. Accumulate enough of these genetic changes in one population of a particular species, and that population could evolve into a new species. This process, whereby members of one species become another species, is called *speciation,* and it's one of the most fascinating areas of evolutionary biology.

As you can imagine, speciation can take a very long time — at least compared with the human life span. For that reason, scientists can't perform a single experiment in a laboratory that allows them to start with one species and watch a new one evolve. What they can do instead is observe the process in nature and study the individual parts of the process in the laboratory and in the natural environment.

Evolution doesn't need to lead to speciation. A species can evolve through time without splitting to create a second species. The study of speciation is the study of what factors are responsible for causing one species split into two species.

## Species and Speciation at a Glance

*Speciation* is simply the process whereby a single lineage splits into two lineages. In other words, new species arise from existing species.

But what constitutes a species and how does speciation occur? The following sections explain.

## The biological-species concept

Scientists have a pretty good handle on what constitutes a species for sexually reproducing animals: the *biological-species concept.* According to this concept, a species is a group of organisms that can interbreed and produce viable and fertile offspring.

Individuals can mate and reproduce with members of their own species but not with members of other species. The defining characteristic separating one species from another is that they are reproductively isolated from each other. When a speciation event occurs — when evolution results in members of one species developing into another species — that group of individuals can no longer interbreed with members of the original species.

The biological-species concept best applies to sexually reproducing animals. It doesn't adequately define what bacterial species are. In fact, defining the term *species* in other cases is an active area of evolutionary biology. For more on bacterial species, head to the later section "A Species Concept for Bacteria."

# When one species becomes two

When new species arise from existing species, you have speciation. Here's how it works: Two different populations of the same species evolve in different ways. They become progressively more different until they are so different that they are no longer able to interbreed. That's all there is to it.

You've heard about speciation, and it might be one of the major reasons you bought this book. Can it be that such a thing really happens? It's clear that some people don't even like the idea of it. How do we know that the whole idea of speciation wasn't just something that Darwin concocted with after one too many beers? Because of ring species, which are explained in the next section. At this point, suffice it to say that by studying ring species, scientists know that a gradual accumulation of small differences is sufficient to cause two populations of the same species to become reproductively isolated.

Of course, ring species aren't the only interesting thing in this chapter. I hope you read the rest of the chapter, too, because speciation is a fascinating topic and one that evolutionary biologists think about a lot. But even if you read no more than the next section, you'll see how scientists — and now you — know that speciation can indeed happen.

# Going in Circles: Ring Species

Through the existence of ring species, scientists can say with 100 percent certainty that small differences can accumulate in nature to the point that two populations of the same species can become reproductively isolated. They can actually go out and see it.

## Subspecies, races, and breeds

Variation exists among different populations of the same species. The following terms are used to describe the different types of variation:

- **Subspecies:** A group within a particular species that shares genetic characters with other group members that it doesn't share with members of the larger species. Subspecies may interbreed quite freely or may be partially reproductively isolated — that is, they can interbreed but don't do it as well, or produce offspring as viable, as when they mate within their own subspecies group. Subspecies can range from ever so slightly-different groups within a species to groups that are on the verge of speciating.

- **Race:** Used most often to describe variation within the human species. Human races are differentiated primarily by skin color, but even though the genes responsible for skin color are noticeable, the actual genetic differences among races are slight. In fact, skin color doesn't accurately reflect the genetic differences among humans. Two people of African descent could easily be more genetically different from each other than a person of European descent may be from a person of Asian descent. Bottom line: Races have slight differences, and these differences are no where near the level they'd have to be to decrease gene exchange.

- **Breed:** Domestic animals (such as dogs and cows) whose characteristics are artificially selected and maintained by humans through animal husbandry are divided into breeds. The goal of selective breeding is to create animals that differ from their wild counterparts and possess relatively predictable traits. Take dogs, for example. Humans have been breeding dogs for only a relatively short period, and over that time, starting with wolves, we've managed to produce everything from Chihuahuas to Great Danes. All breeds of dogs are the same species. They can all interbreed, although admittedly, interbreeding is easier for some pairs than for others. *Note:* Standard convention gives species names to products of animal husbandry. That doesn't mean, however, that dogs and wolves are different species (in fact they share the same species name: *canis domesticus*). Dogs and wolves can still interbreed, even though a "happily ever after" probably wouldn't be in the cards for the Big Bad Wolf and your Pomeranian.

Ring species are species with these specific features:

- **Their habitat surrounds an area of hostile environment that they can't cross.** Think about a bird species living in the lower elevations around the Himalayan mountain plateau. Or a little salamander living around the edges of California Central Valley. Or a bird that lives on the land masses surrounding the Arctic ice cap. They can move around the edges, but they can't cross over the middle, where they wouldn't be able to survive. Figure 8-1 shows a habitat for a ring species.

- **Neighboring subpopulations around the circle, or ring, are slightly genetically different from each other.** These genetic differences can be measured. Maybe there's been selection for different alleles in different places; maybe the genetic differences are the result of drift; maybe both.

- **Most neighboring populations can interbreed with each other.** The populations near one another are a little different genetically, but they are still the same species and can therefore mate and produce viable offspring.

- **At one place around the ring — the ends — the neighboring populations can't interbreed with each other.** Each population can breed with its neighbors (because neighboring populations are just a little bit different), yet all those differences add up as you go from one end of the ring to the other. The result is that, by the time you've gone all the way around from the beginning of the circle to the end, the two populations on the ends are too different to interbreed. If it wasn't for all the populations in the middle, the two end populations would be different species.

Think of the ring as a horse shoe with the ends bent together so they touch. The two populations at the ends of the horse shoe are just too different to interbreed.

San Joaquin Valley, CA

**Figure 8-1:**
The range
for a ring
species.

You can still think of populations comprising a ring species as being part of the same species because they share a common gene pool, and their genes can be combined via the intermediate populations (so they're not completely reproductively isolated), but obviously, this is a gray area because the populations at the ends of the range can't interbreed.

It wouldn't take much for a ring species to become two separate species. Nothing evolutionary has to happen (that's happened already): All you need is for something to wipe out some of the middle populations and break the ring. What could cause the elimination of the middle populations? Take your pick:

✔ Any event, natural or manmade, leading to a fractured habitat that literally separates the range (such as an erupting volcano or earthquake)

✔ Any event that wipes out the populations in the middle (such as an epidemic or uncontrolled hunting).

# The Components of Speciation

Speciation usually doesn't happen overnight. It's a gradual process that involves these 3 components:

✔ **Reduction in *gene flow* (the exchange of genes between adjacent, or nearby populations):** The mating and reproduction that go on within a species does a pretty effective job of keeping all the genes between populations mixed and, therefore, keeping these populations genetically homogenous. As long as the genes can be easily exchanged between populations, the two populations can't diverge genetically. Mating keeps mixing the genes back together. For speciation to occur there needs to be a reduction in gene flow.

✔ **A decrease in the genetic similarities among populations within a species:** Once populations aren't being mixed together, they can become more dissimilar as a consequence of the different evolutionary trajectories experienced by the two populations.

✔ **The development of reproductive isolation between the two populations:** Reproductive isolation can happen via two mechanisms.

 • The accumulation of differences can, by itself, lead to a reduction in the ability to interbreed. The different populations are just too different.

 • When the populations have diverged to the extent that the offspring of such matings are less fit, natural selection acts to prevent mating between individuals of the two populations. If individuals from the two populations can still interbreed but with reduced success, natural selection will favor individuals that say, "I'm just not that into you." Matings that won't produce quality offspring aren't good for fitness — alleles for avoiding such mistakes will increase as a result of selection.

Numerous events can bring about reduction in gene flow, decrease in similarities among populations, and their eventual reproductive isolation. Sometimes, these events are physical barriers; other times, they're not.

## How little changes add up: Local adaptations

As Chapter 4 explains, genetic variation occurs among members of the same species — hardly an earth-shattering piece of information. Just look around. All people are humans, but identical twins aside, all of us look different. (Even identical twins aren't exactly identical, of course, because of environmental factors that affect their phenotypes; refer to Chapter 4.)

By studying naturally occurring variation, scientists have discovered that some of this variation can be accounted for by organisms having adapted to their local environment. Think about it: Some species have ranges that extend over areas so large that environmental factors differ from place to place within the species' range. A species of flowering plant, for example, can have a range that extends from Texas to Minnesota, and given the very different conditions in the two areas, it shouldn't come as a surprise that plants in Texas and Minnesota would flower at different times. These differences often have a genetic basis; natural selection has favored alleles for flowering earlier in Texas and later in Minnesota.

If you're a flowering plant, you want to flower at the same time that your neighbors are flowering, because mating is hard if your flowers are the only ones that are open. For that reason, flowering time, which is crucial to a plant's fitness, is under strong selection. A plant that flowers too early may lose its flowers to a late frost; one that flowers too late may lack sufficient time to produce seeds before winter comes.

A fair number of plants self-pollinate — information that is of absolutely no value in this discussion except as a little bit of trivia. Mull it over at your leisure.

One of the tricks plants use to make sure that they flower at the right time is to assess day length — a perfect predictor of the time of year regardless of where you are. The "right" day-night cycle for flowering, though, depends on where the plant grows. By taking sample plants of a species from different places across the entire species range and growing them under controlled conditions in a greenhouse where day length can be varied, scientists have found that the various populations of plants have adapted to respond to the different day-night cycles of their respective original locations, even though they all belong to the same species.

Although plants with different light-dark cycles can interbreed, they generally don't, for the simple reason that the plants that are most different are likely to be ones that are farthest away, and Minnesota pollen has little chance of landing on a Texas flower. Because *gene flow,* gene exchange between adjacent populations, is low to non-existent over such long distances, plants in different parts of this large area can evolve different combinations of genes. By growing the plants from different locations in a controlled environment, scientists were able to show that some populations in the same species are adapted to the local conditions where they were found.

# Reproductive isolation: The final step of speciation

Speciation occurs when the separate populations become reproductively isolated, losing their ability to produce live or fertile offspring. The mechanisms responsible for reproductive incompatibility fall into two categories: prezygotic isolating mechanisms and postzygotic isolating mechanisms.

- **Prezygotic isolating mechanisms:** Specifically, *prezygotic* refers to those things that occur before the egg is fertilized. Basically, this type of isolating mechanism stops sperm and egg from getting together.

- **Postzygotic isolating mechanisms:** Postzygotic isolating mechanisms are those that occur after the egg is fertilized. They don't stop mating, but they stop the offspring from being viable or able to reproduce.

The following sections go into more detail.

### Prezygotic isolating mechanisms

Anything that could prevent the sperm and egg from coming together is considered to be a prezygotic isolating mechanism, such as the following:

- **Reproductive timing differences:** Hey, when the time isn't right, what can you do?

- **Spatial separation:** If one population is always on hawthorn trees, and the other is on apple trees, never the twain shall meet.

- **Mate-choice specificity:** Basically, this situation is the evolution of being picky.

- **Physical incompatibilities between the two sexes:** This mechanism is common among insects. Have you ever seen *Drosophila* genitalia? Let me just say this: If the pieces don't fit, you really must quit.

- **Inability of the sperm and egg to fuse:** No fusing means no offspring.

### Postzygotic isolating mechanisms

Postzygotic isolating mechanisms are those that come into play even though individuals from the two diverging groups do mate. As a result of different evolutionary trajectories, the parents are different enough that they don't produce fertile offspring (or any offspring, perhaps). Postzygotic isolating mechanisms include

- **Spontaneous abortion of hybrid embryos:** The offspring are never born.
- **Low offspring viability:** The offspring die, often before reproducing.
- **Offspring sterility:** The offspring themselves can't reproduce. (As far as one's fitness goes, producing sterile offspring is exactly the same as producing no offspring at all.)

Postzygotic isolating mechanisms drive the evolution of prezygotic isolating mechanisms. Why? Because mating with someone you can't produce viable offspring with is a bad idea, and natural selection favors genes connected with correct mate choice.

# Types of Speciation

The ring species example is all we need to be sure that the process of speciation can happen. There's nothing magical about: Little differences add up until you get a big difference. But most species don't have a ring-shaped distributions, so while ring species provide an excellent example of the nuts and bolts of the process, that's not how we usually imagine the speciation happening.

## Allopatric speciation: There IS a mountain high enough

In *allopatric* speciation, a physical barrier separates the populations and limits or eliminates gene flow between them. Any physical barrier that reduces gene flow will do: a mountain range, a patch of unsuitable habitat that's difficult to cross, or an ocean.

Allopatric speciation is considered to be the most important speciation mechanism — lots of examples occur in nature — and also the easiest to understand. It just stands to reason that a reduction in gene flow will occur when populations are physically separated, and you can easily see how species can become separated: continents drift apart; mountain ranges rise; climates change and alter the availability of suitable habitat.

Although scientists can't experimentally witness the process of speciation from beginning to end because of the long time periods involved for full speciation to occur, they can find pairs of populations at every stage of the process, from recently physically separated and not very divergent all the way through very divergent and not very able to reproduce.

## Allopatric speciation by founder effect: Getting carried away

Allopatric speciation by founder effect, like allopatric speciation (explained in the preceding section), requires that the two populations be isolated physically. The difference is that in founder effect speciation, the second population originates from a small group of individuals separating from the main group, such as a flock of birds being blown to an island.

Common locations for allopatric speciation by founder effect are islands, which tend to be colonized initially by small numbers of individuals. Scientists find that isolated populations on islands often are very different from the larger mainland population. The farther an island is from the mainland, for example, the less likely it is that large numbers of individuals will be blowing or drifting there. Hence, the differences between the mainland and island populations can be attributed to either (or both) of the following situations:

- ✔ **Genetic drift**: The founding population already starts off a little differently genetically and thus by chance may have gene combinations that predispose it to a different evolutionary future. This process may be helped by genetic drift in the small initial population. For the details on genetic drift, go to Chapter 6.

- ✔ **Natural selection**: When a population is in a novel environment, genetic differences accumulate as a result of natural selection. Differences in predators, prey, or other food resources could drive natural selection in different directions compared with the mainland population.

## Parapatric speciation: I just can't live in your world

In *parapatric* speciation, the two populations aren't physically separated; instead, they abut each other. Because they're within mating distance, something other than a physical barrier must be causing the reduction in gene flow between these populations. That impediment? Natural selection.

Suppose that a species' range encompasses two adjacent environments with conditions different enough that genes advantageous on one side aren't advantageous on the other. As a consequence, natural selection favors particular traits on one side that it doesn't favor on the other side. If an individual from one side meanders over to the other side for a little procreation, his offspring with his genes from the "bad" side will be selected against. As a result, the usual homogenizing effect of mating between individuals of two habitats will be reduced or eliminated, and the two populations will become more divergent.

A classic example of the first stages of parapatric population divergence is the evolution of plants in areas contaminated with mine waste such as heavy metals. These areas aren't ideal places to set up house, but some plants can. In these hostile environments, natural selection favors genes that allow plants to survive high concentrations of lead and other metals. In the absence of this contamination, these genes aren't necessarily favored.

These plants haven't speciated yet; some interbreeding still occurs. Interestingly, the hybrid plants — those that form as a result of matings between the mine-waste plants and the native plants — do less well in both the mine-waste environment and the native environment. Therefore, selection favors genes that reduce the likelihood of matings between plants on either side of the line, and as a result, the plants living on the mine waste have changed in a couple of interesting ways:

✓ **They have evolved different flowering times from the original population.** Different flowering times prevent gene flow across the mine waste. The difference in flowering times is most pronounced at the border between the two environments, because plants that are farther from the mine waste are less likely to receive pollen from mine-waste plants and, therefore, are less likely to experience a selective force favoring different flowering time.

✓ **They have evolved to have a higher rate of self-fertilization.** Self-fertilization could be selectively advantageous for two reasons:

• It prevents the production of offspring with low fitness by preventing matings with plants from uncontaminated soil.

• It comes in handy if getting pollen from another compatible plant is difficult.

In parapatric speciation, selection is strong enough to reduce the likelihood that genes from one environment will make it in the gene pool of a population living in a different environment. Further evolution of reproductive characters (like different flowering times, for example) decreases gene flow between populations even more. In the mine-waste example, the two populations are adjacent, and pollen blows back and forth, yet the offspring that result from matings between the two populations are not likely to contribute their genes to the next generation. Different selective regimes in the two environments have resulted in a reduction in gene flow between the two populations, which will only increase the degree to which the two populations diverge.

# Sympatric speciation: Let's just be friends

*Sympatric* speciation occurs without the organisms in the two populations being physically separated at all. No physical barrier prevents gene flow (as in allopatric or allopatric speciation by founder effect speciation), and no spatial discontinuity exists (as in parapatric speciation). In sympatric speciation, it is some detail of the environment that results in a reduction of gene flow between the two populations.

Suppose that one particular combination of genes makes some individuals better at foraging at night, and another combination generates individuals that are better at foraging during the day. If some of the individuals are active only at night and others are active only during the day, the night-active critters are more likely to breed with other night-active critters, simply because those are the creatures they interact with.

Although researchers suspect that sympatric speciation is unlikely to be a very common speciation mechanism, evidence suggests that the process can be important occasionally. A possible example is the case of the apple maggot worm — a pest of apples in the United States.

Apples aren't native to the United States; they were introduced by European settlers. The apple maggot worm, on the other hand, is native to the United States. Before the introduction of apples, it fed on hawthorns, and many maggot worms still feed on hawthorns. Hawthorns and apples can grow in similar locations. The apple worms that infest apple trees have ample opportunity to mate with worms from hawthorns and produce offspring, but when scientists examine them, they see that the apple worms infesting one type of tree are genetically different from those feeding on the other type of tree.

Something's preventing gene flow between these two types of worms. Even though apples and hawthorns exist in the same environment, for some reason the flies that lay eggs on one species avoid laying eggs on the other. One possible answer involves ripening times. Apples ripen earlier than hawthorn fruits, and the flies living on the different species reproduce at slightly different times.

More interesting, the flies seem to have developed a preference for either apples or hawthorns. The flies laying eggs on apples have a preference for apples, whereas the flies laying eggs on hawthorns have a preference for hawthorns — preferences that must have evolved after the introduction of apples to the United States, arising in a location where apples and hawthorns coexist. It's this feeding preference that's driving the flies' divergence.

Speciation hasn't occurred for the apple maggot flies yet. The two populations can still reproduce (even though doing so would be the entomological equivalent of a Capulet-Montague pairing), but they're beginning to diverge genetically even though they live in exactly the same place, as close to each other as neighboring trees.

CASE STUDY

# Diverging on the fly

As two populations diverge, they can become reproductively isolated. Populations living in different environments experience different selective pressures. As a result, they become genetically different and less able to reproduce. In this case, reproductive isolation is a by-product of selection. Alternatively, reproductive isolation can be selected for when two populations have diverged to the point where hybrids have lower fitness. In that case, any mutations that result in the organisms' wanting to avoid "interbreeding" will increase in frequency.

Diane Dodd designed an experiment in which several generations of fruit flies were raised on different kinds of food: either a starch-based food or a sugar maltose-based food. Starting with a single population of flies, the researchers produced different fly lineages. The initial flies were the same, and the experiment was set up in such a way that the only selective pressure was the food type. Flies that did well on maltose (or starch) were more likely to leave more descendants and hence their genes were more likely to end up in the next maltose (or starch) generation. After 8 generations, Dodd conducted experiments to investigate the mating preferences of the evolved flies.

To conduct these experiments, Dodd introduced male and female flies from the different populations into a cage and kept track of which flies got together. When the flies came from different populations that had been fed the same food, they didn't exhibit a preference for mating with flies from their own population. A fly raised on starch was just as likely to mate with a fly raised on starch from a different population as it was to mate with a fly raised on starch from its own population. The same results held for flies raised on maltose: They didn't exhibit a preference for their own population compared with others raised on similar food.

When the experiment involved flies that had been fed different food types, however, the situation changed. In this case, the flies exhibited a distinct preference for mating with flies that had been raised on the same food they had.

What's important about this experiment is that in producing the flies, Dodd didn't select for reproductive preferences; she selected for increased fitness on one type of fly food or another. But as a result of evolutionary changes for increased fitness, the flies also developed changes in mating preference! This experiment clearly indicates that selection for fitness in different environments may result in reduced ability to interbreed.

# Islands: Good Places to Vacation and Speciate

Some places are just plain better for speciating and studying speciation than others — islands, for example. As stated earlier, islands are just tailor-made for allopatric speciation via founder events.

Islands can be hot spots for speciation. Hawaii is a good example. So are the Galapagos Islands, where Darwin got his first insights into the process of speciation by natural selection. Two factors combine to make islands areas that can facilitate speciation:

- ✔ **Isolation:** Islands are by definition places where populations are isolated from the rest of the species. As gene flow is low, opportunities for genetic diversification increase.

- ✔ **Potential for subsequent speciation:** After a species has evolved on one island in a chain, some of its members can blow or drift to another island, where they may diverge further. In the future, some of them may be blown back to the first island — where, if they're different enough, they may diverge from the population on the original island to produce a second new species.

Islands aren't populated solely by species that just happen to wash up on their shores or get dumped there by a wayward breeze. If an archipelago is far enough from the mainland to make colonization from mainland organisms unlikely, the island is often populated by the species that are unique to the island. You can safely assume that these species arose via speciation on the island. Hawaii, for example, has several native species (like ukulele players) that don't exist anywhere else. They arose in the Hawaiian archipelago.

# A Species Concept for Bacteria

The biological-species concept, explained in an earlier section, classifies species based on their ability to mate and reproduce. That system is all well and good for organisms that mate and reproduce, but it leaves lots of other organisms out in the cold.

Bacteria are perhaps the best examples of organisms that reproduce without mating. They simply divide into two daughter bacteria, each of which goes happily on its way, dividing further. Mutations can occur in this process, just as in any other DNA-replication process; hence, bacteria evolve.

### Bacteria categories

When microbiologists go out to the environment (such as some exotic foreign location the likes of which you see on the Discovery Channel or the back of your throat), they find that they can group the microbes they collect into different categories such as:

- ✓ *Escherichia coli (E. coli):* Common gut bacteria
- ✓ *Staphylococcus aureus:* microbes that can cause staph infections
- ✓ *Neisseria meningitides:* One of the many organisms that can cause meningitis

Because scientists can group bacteria into separate, recognizable categories, they give those groups names like *E. coli* and call them species. But the bacterial species doesn't mean the same thing as it does for animals, plants, and other sexually reproducing organisms.

*E. coli* isn't a species because all *E. coli* mate with one another and not with *Neisseria.* It's a "species" for some other reason. The key is determining what cohesive force differentiates one species of bacteria from another. What keeps the groups separate?

### Periodic selection and selective sweeps

One possible cohesive force that could be responsible for the existence of groups of nonsexually reproducing but similar organisms is periodic selection. In *periodic selection,* natural selection favors a mutation that confers high fitness, which leads to a purging of genetic diversity within a group of nonsexual organisms.

The process works this way. Think about our friend *E. coli.* As one *E. coli* bacterium divides, different mutations begin to accumulate, and the population of *E. coli* bacteria gets more and more variable. Now imagine that a mutation arises that is especially beneficial for an *E. coli* bacterium, enabling it to outcompete all the other *E. coli.* It's more fit, and as it takes over, all the other *E. coli* bacteria are eliminated; as a consequence, genetic diversity within the species is reduced.

This process, called a *selective sweep,* may be the reason why scientists can identify species of bacteria and name them based on their overall similarity — something called *phenetic species* or *ecological species.* In this type of system, natural selection periodically reduces the variation found in a species and keeps, for example, all *E. coli* bacteria looking pretty much the same.

# Chapter 9

# Phylogenetics: Reconstructing the Tree of Life

*E*volution can lead to *speciation,* in which two species arise from a single parent species (refer to Chapter 8). Starting with one life form a long, long time ago, the process of evolution has generated the diversity of life forms we see all around us. This process, all by itself, is just ridiculously cool!

Over time, the sum of these speciation events generates what scientists refer to as the *tree of life.* The neat thing about trees of life (scientific name: *phylogenetic trees*) is that they enable us to trace the history of species in much the same way that genealogical trees let people trace their family histories. In a nutshell, phylogenetics lets biologists figure out the actual history of branching (speciation) for a given set of species.

This chapter introduces you to phylogenetic trees, describing what they are and how they're made. Although knowing the history of species is pretty darned amazing in and of itself, these trees can provide a wealth of other information, too — and you also find that info in this chapter.

# Understanding the Importance of Phylogenetic Classification

Scientists in general like to classify things, and they've been doing it for centuries. Why? Not because it makes them feel good (although it does), but because it helps keep things neat while at the same time providing a wealth of information. The simple process of sorting reveals patterns and relationships, and gives clues to past events — revelations that are absolutely fundamental to the study of evolution. This information is so important to evolutionary biologists, in fact, that they've come up with a way to show how related organisms are, evolution-wise, and to classify them accordingly. This method is called *phylogenetic classification.*

The advantage of a phylogenetic classification is that it shows the underlying biological processes that are responsible for the diversity of organisms. Through phylogenetics, scientists have been able to trace the genetic history of different species and, in doing so, have proved that the process of speciation — whereby ancestral species gives rise to descendent species — is real. (For more information on speciation, head to Chapter 8.) In fact, they've shown, as far as available data allows, that all species existing today descended from a single common ancestor.

# Other classification systems, courtesy of Aristotle and Linneaus

The field of biological classification began with Aristotle (384–322 BCE). He concerned himself primarily with the classification of animals and recognized two major groups: those with blood and those without blood. In the "with blood" group, Aristotle recognized five subgroups: birds, things with four legs that lay eggs, things with legs that don't lay eggs, fish, and whales. The things-with-legs-that-lay-eggs group included animals such as crocodiles, lizards, frogs. The things-with-legs-that-don't-lay-eggs group corresponded mostly with what we now call mammals, except that Aristotle's group didn't include whales. (He recognized that whales are different from fish but didn't realize that they're mammals.) The main thing to know about Aristotle's system is that it used nested groups; he divided organisms into two main groups and then created subdivisions within those groups. This structure set the standard for later classification systems.

Carolus Linnaeus (1707–1778) developed Aristotle's idea of a nested classification scheme more completely. He divided all of life into two kingdoms: animal and vegetable. Then he subdivided those kingdoms into classes, divided classes into orders, divided orders into *genera* (singular: *genus*), and divided genera into species. Linnaeus brought a huge amount of order to the study of biology. The diversity of life is far more manageable if it can be broken down into smaller groups, and Linnaeus was the guy who really got the ball rolling on that front.

Today, scientists have modified the Linnean classification system to incorporate new discoveries and understanding. Instead of Linnaeus's two kingdoms (plants and animals), scientists have proposed additional kingdoms corresponding to such things as single-celled organisms and fungi, and even "grab bag" kingdoms for organisms that don't fit into one of the other kingdoms. The modern version of the Linnean classification system, which scientists use today and is probably the one you learned in high school, looks like this:

- Domain
- Kingdom
- Phylum
- Class
- Order
- Family
- Genus
- Species

Classify humans according to this system and you get

- **Domain:** Eukarya (organisms with one or more cells with a nucleus)

- **Kingdom:** Animalia (other kingdoms are plants and fungi)

- **Phylum:** Chordata (animals having a dorsal nerve tube, like the one that runs down your spine)

- **Class:** Mammalia (animals that have hair, nurse their young, and so on. )

- **Order:** Primates (yes, you're a primate — so are orangutans, apes, chimps, and others)

- **Family:** Hominidae (modern man and extinct ancestors of man)

- **Genus:** Homo (species of humans, both extinct and currently living)

- **Species:** sapien (wise — proving scientists do have senses of humor)

Here's a little tip: A fun way to remember each level of this system in order is to use the mnemonic "Did King Phillip Come Over For Good Sex?" And if you leave out Domain, which some systems do, the mnemonic changes only slightly: "King Phillip Came Over For Good Sex."

The evidence that all life descends from a single common ancestor includes such things as the unity of the genetic code. Organisms use a simple code to determine how to make proteins from DNA sequences, and all organisms use the same code (although some minor exceptions exist; go to Chapter 15 for details.) To find out more about the genetic code and its importance to evolution, refer to Chapter 3.

Beyond enabling scientists to trace genetic connections back through time, phylogenetics lets scientists better predict what's to come. Being able to anticipate future mutations is an especially important function in areas like health care; virologists and epidemiologists use info gleaned from phylogenetics to stay one step ahead of the bugs that are trying to stay one step ahead of the human immune system. (You can read more about viruses and the race for vaccines in Chapter 19.)

# Drawing the Tree of Life: Branching Patterns and Speciation

Scientists show phylogenetic relationships by drawing phylogenetic trees. If you trace the process of speciation on paper, you end up with a branching pattern that's referred to as the tree of life. Each branch in the tree represents a speciation event, when one species evolved into another.

## A simple tree

The easiest way to explain phylogenetic trees is to start with the simplest tree possible: one with three species, which I'll call species A, B, and C.

Imagine that you know that species B and C are more closely related to each other than either of them is to species A. More specifically, species B and C have a common ancestor that they don't share with species A. All three species also have a common ancestor that dates from a time before the common ancestor of species B and C. A phylogenetic tree for these species would look like the one shown on the left in Figure 9-1. Plug in species names, and you have the tree shown on the right.

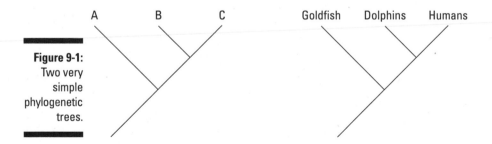

**Figure 9-1:** Two very simple phylogenetic trees.

## A more complex tree

Most phylogenetic trees are not as simple as the one in Figure 9-1, of course. A more realistic (and complex) tree appears in Figure 9-2. This tree shows the relationships among some of the major vertebrate groups. Located at the tips of the branches are the names of the groups (lampreys, sharks, sturgeons, mammals, birds, and so on).

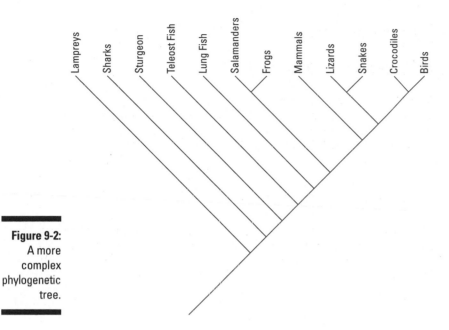

**Figure 9-2:**
A more complex phylogenetic tree.

As stated earlier, Figure 9-2 shows the phylogeny, or evolutionary relatedness, among some of the major vertebrate groups. Strictly speaking, however, the scientific convention would be to refer to this diagram as a *phylogenetic hypothesis* rather than a phylogeny. Not having been around that many millions of years ago when these relationships were forming, modern scientists can only infer how these things evolved; they can't state it directly.

# Reading Trees

Phylogenetic trees convey quite a bit of information. To figure out what that information is, you have to be able to read the tree — that is, to understand the relationships that the tree illustrates. This section explains what you need to know.

## Knowing your nodes

A phylogenetic tree is comprised of branches and *nodes* — places where branches connect — that represent ancestral species (species that give rise to the species at the tips of the branches). Figure 9-3 shows the same tree as does Figure 9-2, except that the nodes are circled. Note that, among the many connections, you can see that crocodiles and birds share a common ancestor. Go a bit farther back in time, and you can see another ancestral species that they share with lizards and snakes.

Scientists refer to nodes as *taxa*. All the organisms that appear at the tips of the tree (in Figure 9-3, the birds, crocodiles, lizards, snakes, mammals, etc.) are *terminal taxa*.

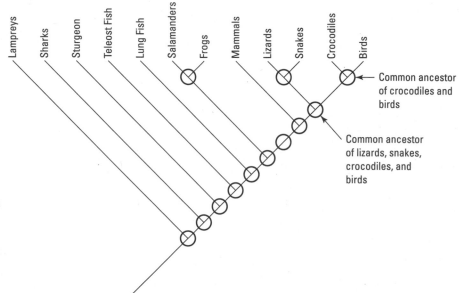

**Figure 9-3:**
Nodes represent ancestral species.

# Getting oriented: Up, down, or round and round

As you read phylogenetic trees, keep in mind that the important thing is the positions of the groups relative to one another. Changes in how the diagram is drawn that don't change these relative positions are unimportant. Exactly the same information is conveyed either way.

Figure 9-4 shows the simple human-dolphin-goldfish tree from Figure 9-1 drawn four different ways, but — and this part is the important part — all four diagrams in this figure represent exactly the same relationships among these organisms. In each diagram, dolphins and humans have a common ancestor more recent than the ancestor they share with goldfish. The differences among the four trees can be described as the results of rotating the branches around the nodes.

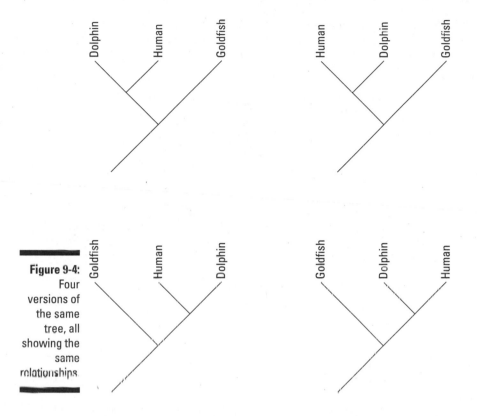

**Figure 9-4:**
Four
versions of
the same
tree, all
showing the
same
relationships.

From each of the diagrams in this figure, you can see that at some point in the past, a speciation event led to the goldfish lineage and to the lineage that subsequently diverged into dolphins and humans. Note that the dolphin and human (both of which are mammals and have a fair bit in common) share a most recent common ancestor that neither shares with the goldfish. But all three species have more distant common ancestor: All three are vertebrates.

The left-to-right order of the diagram conveys *absolutely no information*. You can draw exactly the same tree with humans in the left position or the middle position. The branching pattern relative to the nodes is what's important.

Figure 9-5 shows another, more complex tree with a section that's been rotated. Again, the branches for a particular node can be rotated around a node without changing the information in the tree. Both sections convey exactly the same evolutionary relationships: Lizards and sharks are just as related to each other as they were before, and the combined group of mammals/lizards/sharks/crocodiles/birds is in exactly the same position relative to the other branches of the tree as it was before the rotation. The horizontal order of the tips of the tree is different, but the positions of the groups relative to one another haven't changed.

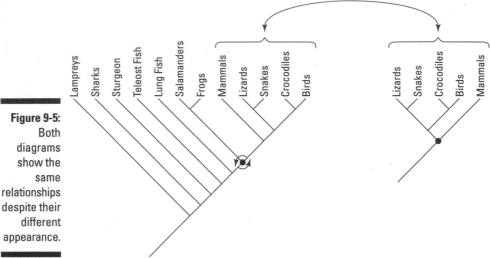

**Figure 9-5:**
Both diagrams show the same relationships despite their different appearance.

# Understanding groups

When you look at a tree, one of the key things you notice is that the branching pattern creates groupings of species. This section explains the two types of groups: monophyletic and paraphyletic:

- **Monophyletic groups** represent all the descendants of a common ancestor.

- **Paraphyletic groups** also represent shared ancestry, but of only part of the group.

Why do we even care about these groups? Because they're "real" — they indicate an actual past connection. Mammals are a monophyletic group, for example, and so are bats. The members may have diverged in different ways, but they all started the same.

## Monophyletic group

A *monophyletic group* is a group of species that (1) has a common ancestor and (2) includes all the descendants of that ancestor. Figure 9-6, for example, includes three monophyletic groups:

- All the *vertebrates* in Figure 9-6 have a common ancestor, indicated by the number 1.

- All the *mammals* comprise a monophyletic group (the bigger box) that includes all the descendants of the common ancestor indicated by the number 2.

- Dolphins and chimps comprise a monophyletic group (the smaller box) descended from the ancestor indicated by the number 3.

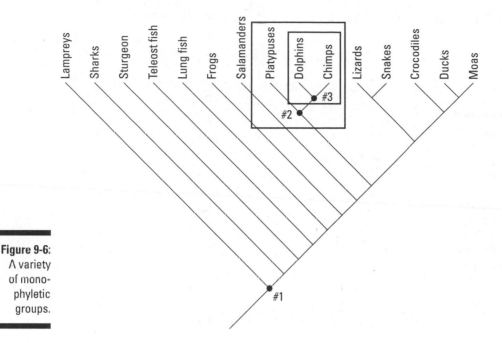

**Figure 9-6:**
A variety
of mono-
phyletic
groups.

A monophyletic group is also referred to as a *clade,* as in "Aren't you clade a monophyletic group has another, equally hard-to-remember name?"

Monophyletic groups are important in a phylogenetic classification system because they're based on the evolutionary process, not on some arbitrarily selected character.

### Paraphyletic group

A group of organisms that has a common ancestor but doesn't include all the descendents of that ancestor is called a *paraphyletic group.* Figure 9-7 shows two paraphyletic groups, the first of which is fish. Although the figure includes several monophyletic groups of fish (sturgeon, lungfish, and teleost), fish as a whole don't make up a monophyletic group. The common ancestor of the fish, indicated by the number 1, gave rise to many other vertebrate groups.

Another paraphyletic group in the vertebrates is the group commonly referred to as the reptiles, represented in Figure 9-7 by the snake, lizard, and crocodile. The reptiles are a paraphyletic group rather than a monophyletic group because the common ancestor of snakes, lizards, and crocodiles also gave rise to birds.

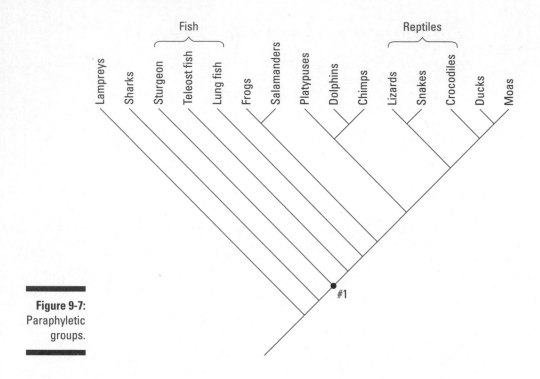

**Figure 9-7:**
Paraphyletic
groups.

# Reconstructing Trees: A How To Guide

Reconstructing a phylogenetic tree involves searching for the clues that tell you about the relationships among different species and then applying rigorous, and often quite complicated, analytical techniques to turn your pile of clues into your best hypothesis about the actual tree.

This process boils down to essentially three steps:

1. **Identify and analyze characters shared by the species for which you're constructing the tree.**

2. **Use outgroup analysis to determine whether each character state is either ancestral or derived.**

3. **Group the species based on your analysis.**

The following sections explain these steps in more detail.

# Finding clues (aka characters)

So what are these clues that you use to make the trees? Well, evolution involves change, and in tree reconstruction, you use these changes to map out the history of evolution. The clues to look for are *changes in the states of characters*. Gathering clues involves

- **Identifying the characters:** *Character* simply refers to an organism's specific, measurable traits. A character can be just about anything — hair, eye color, ability to digest milk, and resistance to antibiotics, for example.

- **Determining the different states the characters have:** The character states could be hair present and hair absent.

- **Polarizing the characters:** *Polarizing* characters refers to determining the direction of evolution with respect to each particular change. When you take all your clues and put them together into a tree, it helps to know which of the character states is the *ancestral state* (the one before the evolutionary event) and which is the *derived state* (the one after the evolutionary event).

Imagine three species — turtle, platypus, and rabbit — for which you want to reconstruct the evolutionary history. First, you find some clues — things about these organisms that allow you to group them based on their ancestry. You decide to start with just two clues:

- Both the platypus and the rabbit have hair.
- Both the platypus and the turtle lay eggs

You know that both characters (hair and egg laying) have two states: present and absent, as in "has hair/doesn't have hair" and "lays eggs/doesn't lay eggs."

When scientists decide to reconstruct a phylogenetic tree, they have to decide what characters, or traits, to use. These characters can be either morphological (visible physical characters) or molecular (genetic). In this section, I limit the discussion to morphological characters, such as wings, flippers, and eggs. Before scientists knew about DNA and DNA sequencing, they were limited to these types of characters as well. Today, the ever-increasing amount of information available about DNA sequences gives scientists the opportunity to use the organism's nucleotide sequence (basically, its genes; refer to Chapter 3) as a character.

## Using outgroup analysis to determine derived and ancestral states

Now you need to determine which character state is ancestral and which is derived. To do that, you use *outgroup analysis*. Collectively, the species you're trying to make a tree for — in this example, the platypus, turtle, and rabbit — are called the *in group*. The *outgroup* includes species that don't belong in the in group. In this example, they'd be things like clams, bumblebees, and mushrooms — things without backbones and four legs and stuff like that.

Although clams, bumblebees, and mushrooms share a common ancestor with rabbits, turtles, and platypuses, that ancestor is pretty far back in the tree. So you can use the character states of these other organisms to assess character polarity, which refers to which character state is the derived one and which is ancestral. Mushrooms aren't much help in this case; they're just too far out. But you can learn some things from other animals:

✔ Almost all the animals in the outgroup lay eggs. From this character, you can conclude that live birth is the derived character state and that laying eggs is ancestral.

✔ None of the organisms in the outgroup has hair. From this character, you can conclude that hair is a derived state.

After you've polarized the characters, you're able to conclude that the split between the platypus and the rabbit occurred *after* the split between the lineage that led to the turtle and the lineage that led to the ancestor of the platypus and the rabbit.

## Grouping species

Based on analyzing the outgroups, you know that "laying eggs" is the ancestral state and "live birth" is the derived state. You also know that "no hair" is the ancestral state and "hair" is the derived state.

From this analysis, you know that in the past, farther down the tree of life, critters laid eggs but didn't have hair. Under this scenario, the change from egg-laying to live birth doesn't help you group any two species as having the most recent common ancestor. Only one of the species — the rabbit — has this derived character state. Because only one species has the derived state, you can't use it to form a group.

The situation is different for hair. Two species — the platypus and the rabbit — have the derived character state, which allows you to group these two species as having a more recent common ancestor than the common ancestor of all three species. You envision that the platypus and the bunny have a hairy ancestor, so they both have hair.

To reconstruct evolutionary history successfully, you need *shared derived characters* — evolutionary events that enable you to differentiate among different groups of organisms. In Figure 9-6, shown earlier in this chapter, the monophyletic-group mammals can be distinguished based on the presence of several novel traits, such as hair and the production of milk. The dolphin and the chimpanzee can be further classified as a monophyletic group based on the derived character of live birth. The platypus retains the ancestral condition of laying eggs.

## Understanding homologous traits

When comparing character states among different species, you have to make certain that you're comparing two different states of the *same* character rather than two completely different characters.

Take the case of forelimbs. Humans have forelimbs; earthworms and bacteria don't. A scan through the diversity of life suggests that having forelimbs is a derived character state. Because the forelimbs in vertebrates share a common evolutionary origin, they are said to be *homologous*.

*Homology,* which is similarity as a result of common descent, helps scientists reconstruct the history of evolution and determine the best estimate of the tree of life. Although forelimbs themselves are homologous, they can evolve in several directions: into wings in birds and bats, flippers in dolphins, arms in humans, front legs in horses, and so on.

## Looking at homoplasies

Traits that are similar for reasons *other* than common history are called *homoplasies*. Homoplasies can have several evolutionary origins, most easily categorized as convergence, parallelism, and reversal:

✔ **Convergence:** Different structures in two different organisms evolve to appear similar. The streamlined shape and fins of dolphins appear similar to those of sharks and fish, but closer analysis reveals that the structures are quite different — an example of convergence. Dolphins and sharks don't share fins because their common ancestor had fins; they share fins because fins evolved independently in the dolphin lineage.

- ✔ **Reversal:** A structure that previously evolved is subsequently lost. The moa is an example of a reversal. Moas, like most of the other vertebrates, don't fly; they're similar to chimpanzees in this regard. But the similarity is not the result of descent from a common ancestor that didn't fly (though it's true that the most recent common ancestor of the moa and chimp didn't have wings). Numerous structural characteristics of the moa skeleton place the moa firmly within the group of birds, and its ancestors had wings, but these wings were lost during the course of evolution to larger body size.

- ✔ **Parallelism:** Two organisms evolve to acquire the same trait, *but* — and this point is key — they don't share this trait because of a common ancestor. Rather, they share the trait because it evolved independently in both lineages. As the preceding section explains, the presence of fore-limbs in birds and bats is the result of homology; both species descended from an ancestor that had front limbs. The fact that the front limbs of bats and birds now function as wings is a *homoplasy* because wings evolved independently in the two different lineages. Most mammals don't fly; the evolution of flight in bats happened independently of the evolution of flight in birds.

The fact that dolphins and sharks evolved fins independently doesn't mean that their common ancestor didn't have fins. Fins are the ancestral condition in vertebrates in the lineage, but they were lost in the lineage leading to the tetrapods and subsequently re-evolved in the lineage leading to the dolphins. (If you feel like your head is going to explode, take heart. Cranial explosion is neither a derived nor an ancestral trait in any known organism.)

The presence of homoplasy throws a wrench into the works, limiting your ability to reconstruct the tree accurately, because it's not always simple to tell when a character similarity is due to homology or homoplasy.

## Testing phylogenetic trees

So how do you know that the phylogenetic trees you can create represent evolutionary relationships accurately? The answer is the way you know anything: Someone did an experiment and checked.

David Hillis and co-researchers set out to determine how well different methods of phylogenetic reconstruction worked at reconstructing the history of evolution. To do this, they needed to know the history of evolution, and they came up with a very clever method.

Taking advantage of the fact that viruses have extremely short generation times, they produced the tree of viral lineages in the laboratory. They started with one virus, from which they collected different mutants. Then they grew these slightly different viruses, collecting mutants from these viruses in turn. In this way they were able to generate a branching tree for which they knew the exact pattern of the branches, because they knew which viruses came out of which flasks.

The different strains of viruses didn't become different species over the course of the experiment (and as Chapter 8 explains, scientists aren't completely clear on what constitutes a species in viruses). But the descendant viruses were different enough from the parent virus to allow Hillis and his co-workers to construct an evolutionary tree.

Viruses don't have a lot of characters like eggs and hair, so the researchers determined what changes had occurred in the viral DNA. They were able to demonstrate that it's possible to measure the traits of the terminal taxa and accurately reconstruct their evolutionary history. In addition, they were able to use the character states of the terminal taxa to make good predictions about the character states of the ancestral taxa at the nodes of the tree. Because they had kept all these ancestral taxa in little vials in a freezer, they were able to show that their predictions were accurate.

# Reconstructing Trees: An Example

One way to reconstruct phylogenetic trees is to use *maximum parsimony analysis.* In this method, you determine the minimum amount of evolution required to explain a particular character set. The tree with the minimum number of evolutionary events is called the *most parsimonious, or shortest,* tree.

You can reconstruct a phylogenetic tree in other ways, such as maximum likelihood, Bayesian analysis, and UPGMA (Unweighted Pair Group Method with Arithmetic mean). Each method has its advantages and disadvantages, but I'm not going to explain them all. I cover parsimony because it involves the smallest amount of mathematics. (Google *maximum likelihood,* and you'll see that you're getting off easy!)

The following sections take you through a simple example of tree construction.

## Identifying characters

Suppose you want to reconstruct a phylogenetic tree for Species A, B, and C, using seven different character states (1 through 7). For this example, what the characters are doesn't matter; the characters could be eggs present or absent, hair present or absent, and so on.

This example is uncomplicated by homoplasies (refer to the earlier section "Looking at homoplasies" for info), but phylogenetic reconstruction can get very complicated very quickly. Fortunately, for the purposes of this book, you don't need to know how to reconstruct complex trees. I just want you to understand the basic process and to realize that reconstructing these trees isn't magic.

## Assigning polarity

Through comparison with the outgroup species (X), which has the ancestral character state for all seven characters under consideration, you assign a character polarity to each one. The number 1 represents the derived character state, and the number 0 represents the ancestral character state.

Looking at the characters for the other three species, suppose that you find the following:

- ✔ For characters 1 and 2, all three species have the derived condition, making species A, B, and C different from species X (the outgroup).
- ✔ For characters 3 and 4, only species B and C have the derived characteristics.
- ✔ For characters 5, 6, and 7, only species C has the derived characters.

## Grouping species

Because species A, B, and C have the derived character for characters 1 and 2, you know that they are a monophyletic group that doesn't include species X. On the phylogenetic tree, you indicate characters 1 and 2 with two slash marks — one labeled with the number 1 and another labeled with the number 2 — at a point below the common ancestor of A, B, and C (see Figure 9-8). It's most parsimonious to assume that these two evolutionary events happened only one time.

Because only species B and C have the derived character states of characters 3 and 4, you separate these two species as a monophyletic group and indicate the characters 3 and 4 with slash marks below the point of *their* common ancestor (see Figure 9-9).

In the case of characters 5, 6, and 7, only species C has the derived characters. Because these derived traits aren't shared with any other species, they don't give you any information about the topology of the tree. (Remember, to group species within a tree, you must have shared characters.)

Placing these characters on the tree adds to the total *tree length,* which is defined as the total number of changes required to explain the data matrix, which in this case is seven (see Figure 9-10).

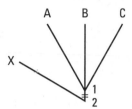

**Characters**

**Figure 9-8:** Character states for different species.

| Taxon | 1 | 2 | 3 | 4 | 5 | 6 | 7 |
|---|---|---|---|---|---|---|---|
| X (outgroup) | 0 | 0 | 0 | 0 | 0 | 0 | 0 |
| A | 1 | 1 | 0 | 0 | 0 | 0 | 0 |
| B | 1 | 1 | 1 | 1 | 0 | 0 | 0 |
| C | 1 | 1 | 1 | 1 | 1 | 1 | 1 |

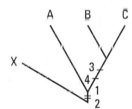

**Figure 9-9:** Species A, B, and C comprise one monophyletic group; species B and C another.

**Characters**

| Taxon | 1 | 2 | 3 | 4 | 5 | 6 | 7 |
|---|---|---|---|---|---|---|---|
| X (outgroup) | 0 | 0 | 0 | 0 | 0 | 0 | 0 |
| A | 1 | 1 | 0 | 0 | 0 | 0 | 0 |
| B | 1 | 1 | 1 | 1 | 0 | 0 | 0 |
| C | 1 | 1 | 1 | 1 | 1 | 1 | 1 |

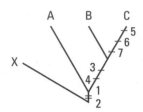

| | Characters | | | | | | |
|---|---|---|---|---|---|---|---|
| **Taxon** | **1** | **2** | **3** | **4** | **5** | **6** | **7** |
| **X (outgroup)** | 0 | 0 | 0 | 0 | 0 | 0 | 0 |
| **A** | 1 | 1 | 0 | 0 | 0 | 0 | 0 |
| **B** | 1 | 1 | 1 | 1 | 0 | 0 | 0 |
| **C** | 1 | 1 | 1 | 1 | 1 | 1 | 1 |

**Figure 9-10:** Only species C has derived characters 5, 6, and 7.

## A word about more complicated trees

The preceding example includes only homologous characters — those that appear as a result of sharing a common ancestor. When homoplasies are involved (similar characters that don't indicate a common ancestor), things get more complicated.

If you assume that sharing similar characteristics means that species belong in monophyletic groups and plot the species accordingly, you end up in with a tree that doesn't make sense: A species may appear in different groups, for example, indicating that they evolved more than once.

To avoid such a scenario, you need to remember that similar characters can evolve independently, or they can appear, disappear in subsequent evolutionary changes, and then reappear. When multiple scenarios are possible, the one(s) you support end up being those that have the fewest (or most parsimonious) evolutionary steps.

# Seeing Phylogenetic Trees in Action

As this chapter shows, it's possible to reconstruct the history of evolution through phylogenetic trees. Now the question is how scientists can use these trees. The answer is that they can use trees for all sorts of purposes, like these:

- Reconstructing the history of human migration patterns
- Quantifying the process of co-evolution

- ✔ Tracing the spread of the human immunodeficiency virus (HIV) epidemic
- ✔ Designing a better flu vaccine

The following paragraphs offer several examples that involve the HIV virus. The other items are covered in detail in other chapters.

# Example 1: The Florida dentist

In the early 1990s, a large number of patients of a particular Florida dentist contracted HIV. The research question at the time was whether they could have contracted the HIV virus from the dentist. Specifically, given that many strains of the HIV virus were circulating in the community, was it possible to connect the HIV strains found in the infected patients with the virus found in their dentist?

By constructing a phylogenetic tree that included the dentist's HIV strain, the six patients' HIV strains, and a selection of HIV strains from the broader community, investigators showed that the patient strains had a recent common ancestor with the dentist's strain, not with the strains sampled from the community.

The take-home message is that even though researchers don't know exactly how these viruses got from the dentist to the patients, they know that the patients did indeed contract HIV from the dentist.

# Example 2: General exposure to HIV

In this example, the question is how did the human species become exposed to HIV? Phylogenetic analysis can address this question by reconstructing an evolutionary tree of not just HIV, but of other species' immunodeficiency viruses as well.

The two major types of human immunodeficiency virus are HIV-1 and HIV-2. Reconstructing the tree of immunodeficiency viruses reveals that HIV-1 strains are most closely related to the simian immunodeficiency virus found in chimpanzees, whereas HIV-2 strains are more closely related to the simian immunodeficiency virus found in sooty mangabeys.

The fact that each human virus is related to a different simian virus indicates that the human viruses are the result of two separate events in which a simian immunodeficiency virus jumped to a human host. The phylogenetic analysis that produced this information also directs researchers' attention to the specific simian viruses that are implicated as the parent of the human infection.

You can read more about the origins of HIV in Chapter 18.

## Example 3: Legal cases

The science of phylogenetics has made its way into the legal world, where it has been used to prove both guilt and innocence. In the first case, a doctor was convicted of attempted murder after infecting his former girlfriend with an HIV strain from one of his patients. This case was the first instance in which phylogenetic evidence was admitted in a U.S. court. Phylogenetic analysis of HIV sequences from the infected woman and the patient, as well as analysis of additional sequences from the community, revealed that the infected woman's HIV strain was most closely related to the strain from the patient. Her strain even had resistance genes against the HIV medications with which the patient was being treated.

The second case involved six foreign health workers in Libya accused of intentionally spreading the HIV virus to hospital patients. An analysis of the HIV strains infecting those patients revealed such a diversity of strains that the parent strain from which the strains evolved would have to have been present in Libya before any of the foreign health workers arrived. Initially sentenced to death despite the scientific evidence supporting their innocence, the sentence was later commuted to life imprisonment. Jailed since 1999, the workers were finally able to leave Libya in the summer of 2007 following diplomatic negotiations.

# Part III
# What Evolution Does

The 5th Wave                    By Rich Tennant

THE FIRST HOMO ERECTUS

"Is this so difficult? Just bend at the waist
and lift with your legs."

# In this part . . .

An organism's genetic make up has a huge impact on more than just its physical characteristics, like how many limbs it has or how fast it can run. It can also affect sex selection (how an organism picks a mate), social behaviors (how — and how well — an organism gets along with others), and life histories (life spans and age of reproductive viability). See the connection? If genes impact everything, and evolution impacts genes, then there's an evolutionary component in not only how we look, but also how we behave and interact. This part explains it all.

# Chapter 10

# The Evolution of Life History

*In This Chapter*

▶ Exploring life histories

▶ Understanding fitness trade-offs and constraints

▶ Appreciating the many ways to be fit

he specific details of an organism's life cycle and reproductive strategy are its *life history,* which includes longevity, age at reproductive maturity, how often the organism reproduces, and how many offspring are produced. Life histories are fascinating to evolutionary biologists because so many of them exist and can be so different. Of course, all this information is interesting at the individual level, but evolutionary scientists look at life histories at the species level.

An oak tree will make thousands of acorns over its life, but it's not likely you're going to make more than one minivan full of children. The human life history is very different from the oak tree life history, yet both strategies have been very good at getting offspring into the next generation. There are lots of people and lots of oak trees, but from a life history perspective, these two species go about it in different ways.

This chapter introduces you to the current theories concerning life-history evolution, theories that seek to answer questions such as these:

✔ Why is it best sometimes to make lots of eggs once, but at other times it's best to make one egg lots of times?

✔ Why do some animals live for weeks and others for decades?

✔ Why do we grow old, and why do we die when we do?

# Evolution and the Diversity of Life Histories

Life histories — the specific details of an organism's life cycle and reproductive strategy — differ greatly among species. To get a glimpse of this diversity, consider the differences between a penguin and a salmon.

A salmon swims up a river to spawn, struggling against the current, jumping waterfalls, and dodging hungry bears. After years of living in the ocean chasing prey, avoiding predators, and storing up enough energy, it makes this final trip to the place of its birth. When it arrives, it produces thousands of eggs and dies soon after.

An emperor penguin — the only terrestrial vertebrate (other than a few scientists) to winter in Antarctica — can live more than 20 years. Each year, it makes a perilous trip to its inland breeding grounds, walking almost 100 miles in tiny steps, only to produce a single egg.

Both of these species have been shaped by evolution, but in very different ways. What it takes to be a fit salmon is obviously very different from what it takes to be a fit penguin. And before you say well, sure, one's a fish and the other's a bird, remember that not all fish die after reproducing once, and not all birds lay one egg a year for 20 years. Life histories vary significantly even among the same types of animals.

Scientists understand enough of the underlying genetics to know that life-history characteristics are heritable and that, over time, they change, or evolve. The fact of life-history evolution is that like evolution of other traits, it happens. What evolutionary biologists want to understand is why. How does selection drive this process? And how can so many different life-history strategies exist?

Scientists have a theoretical framework that explains the facts they observe, and they can test these ideas both in the lab and in nature. Concepts that at first seemed confusing (such as death and aging) make sense now, in light of an understanding of the evolutionary process.

Organisms don't live in a vacuum. The selective pressures they experience are a combination of their *biotic* and *abiotic* environment — that is, the organisms they interact with and the physical factors (temperature and such) they contend with. These factors are different for each organism, and the variations are responsible for the corresponding diversity of life-history patterns. In other words, no single life-history strategy is best because no one pattern could be best across all the different environments. Fortunately, scientists have a good understanding of how specific environmental differences influence the evolution of life histories.

# 'Til Death Do Us Part: The Evolution of Life Span

Evolution via natural selection acts to maximize fitness, and fitness is all about making sure that your genes are around in the future. That being the case, dying just doesn't seem like an especially good idea. After all, death doesn't seem to bode well for the "Get your genes to last forever" imperative of natural selection.

You may think that the simplest way to get your genes into the future is for *you* to exist into the future. After all, you've got all your genes; if you live into the future, your genes live into the future, too. Voilà — fitness without the rather depressing and (often) messy process of dying. But organisms don't get their genes into the future by living forever, even though they may live for a very long time. All this leaves evolutionary biologists, and not just older ones, puzzling over why things eventually (or not so eventually) die.

Evolution has led to many types of life spans. Giant sequoias, for example, live for thousands of years, but most plants have much shorter life spans. Your pet guinea pig, with the best food and care, might live as long as 8 years, but humans live longer than that, and other animals live much longer than we do.

## Why die? Trade-offs and risks

It's not hard to think of genes that definitely should increase in frequency. Imagine an animal that lived forever; reproduced early and often; and had huge numbers of offspring, all of which survived. Talk about fit! Those are some fantastic genes. But these genes don't occur in nature. Why not? Two reasons:

- ✔ **Trade-offs:** Often, one thing happens only at the expense of another.
- ✔ **Risks:** The longer you're around, the more likely it is that something bad will happen to you.

### Farewell, sweet Harriet

Biologists were much saddened in 2006 by the passing of Harriet, age 176 (approximate). Before you run out to get whatever vitamins she must have been taking, I should mention that Harriet was a giant tortoise Charles Darwin collected from the Galapagos Islands.

Tortoise or not, Harriet showed us via her longevity that an animal can live to be 176 years old. The heart can beat that long, brain cells can think that long — it's biologically possible. But most animals (sadly, humans included) just don't bother.

A common misconception is that dying is an adaptation to make room for younger, more vigorous individuals — the idea that organisms die for the good of the species. Although this theory may sound good at first, remember that selection acts most strongly at the level of individuals. In a population in which some individuals have a gene for graciously dying to make room for everybody else and other individuals don't, it wouldn't take long for the "die graciously" gene to go extinct.

Even if such a gene were good for the species as a whole, it would be bad for the individuals that had it because they'd be more likely to die, making them *less* likely to pass this gene on to the next generation. As a result, the gene would decrease in frequency over time and vanish from the population, along with any individuals who were dying for the good of the population. Not surprising, nature shows no evidence of a "die graciously" gene.

### Trade-offs: Evolutionary cost-benefit analysis

The different life-history components involve trade-offs. Because an organism has only a finite number of resources available, it doesn't have the energy to do everything. Energy spent on reproducing, for example, is energy that can't be spent on surviving. Reproducing early and often may mean not having enough energy left to stay alive.

In this scenario, allocating lots of resource for reproducing even though it makes you die sooner would be adaptive if, by trading longer life for more offspring, the organism increases its ability to pass on its genes. The crucial point here is that the organism is getting more copies of its *own* genes into the next generation, not just making room for somebody else's.

### Risky business

Living longer can be risky. The longer an organism lives, the greater the chance that it will become ill or get eaten by a predator. As the risk of death increases, so does the advantage of earlier reproduction, even if this early reproduction results in a shorter life span.

Think of the salmon, which jumps waterfalls and dodges bears to make it all the way upstream: She puts every last calorie into reproducing because the chance of making it up that river twice is too small to make reproducing fewer eggs the first time worth the risk. When the risk of death is higher, natural selection favors genes for earlier reproduction.

# Methuselah flies: The evolution of life span in the laboratory

Laboratory experiments have shown that life-history traits — specifically, life span and metabolic trade-offs — do evolve as scientists expect. This section

## Test tubes and teacups

Peter Brian Medawar was the first person to articulate the idea that allocating resources to survival instead of reproduction increased exposure to risk. Where'd he get this idea? By imagining a population of test tubes. Test tubes don't reproduce, but neither do they grow old and die. So why do you ever have to buy more test tubes? The answer, of course, is that test tubes experience accidental death: They get dropped or knocked over, shattering into pieces so that they're no longer usable.

You can say the same thing about teacups. Even though teacups have been made for thousands of years, you've probably had to buy some yourself, because they didn't all survive.

Medawar's key insight was recognizing that an organism whose strategy for making sure its genes were around in the future consisted of devoting all its energy to surviving instead of reproducing would eventually run out of luck. Even if it's possible to avoid aging, in the end the risk of death remains by means other than old age. You could get eaten, for example, or you could be crushed by a falling tree.

As an interesting aside, Medawar spent only part of his time thinking about evolutionary issues and shopping for test tubes. His primary research involved the immune system — work for which he received the Nobel Prize in Physiology or Medicine in 1960.

looks at one experiment conducted on fruit flies by evolutionary biologist Michael Rose and company. With their fruit flies, these scientists conducted two experiments that tested the following:

- ✓ **Whether aging could be postponed by strengthening the force of selection at later ages.** If so, this result would provide evidence that a contributing factor of aging is that selection doesn't remove mutations that are only harmful late in life, after you're done reproducing. The mutations aren't neutral from an individual fly's health perspective, but they are *selectively neutral* because, by the time they rear their ugly heads, they've already been passed on to the next generation. An experiment that increases the strength of selection at later ages changes these mutations (assuming they exist) from neutral (because there isn't any selection at later ages) to deleterious (because the researchers have added selection at later ages).

- ✓ **Whether metabolic trade-offs may be involved in life-history evolution.** If so, this result would show that the trade-offs are real. It makes sense that energy spent doing one thing can't be spent doing something else. The scientists hypothesized that spending all your energy on reproduction means you have less energy to devote to survival because of these trade-offs.

## The art of testing evolutionary theories

Biologists often perform experiments to test evolutionary theories. People often wonder how this is possible, because they think that evolution is supposed to take a really long time. The answer is evolution can take a long time, but it doesn't have to. The trick is picking the right organisms — ones that live fast and die soon. Insects, for example, can be good experimental subjects, but elephants — not so good.

You need organisms with very short life spans so that you can squeeze in as many generations as possible. These organisms should also be small, so that it's easy to keep large numbers of them, and they should be easy to raise in a laboratory.

Fruit flies (often, the species *Drosophila melanogaster*) meet these criteria nicely.

Laboratory experiments replace natural selection with *artificial selection,* which means that the experimenter — not nature — decides what traits will increase an organism's chance of contributing to the next generation. This replacement is done for many generations, and the researchers track how the organisms change in response to the laboratory selection regime.

Humans have actually been using artificial selection for thousands of years. It's how we make different breeds of dogs, cows that give more milk, and roses that are more resistant to pests.

### Experimental selection for increased life span

To test the hypothesis that aging can be postponed by strengthening the force of selection at later ages, Rose set up 10 replicate populations of fruit flies in his laboratory. For each population, he allowed the flies to feed, mate, and lay eggs; then he transferred a sample of eggs to a new container with fresh food. These eggs hatched; the flies matured, mated, and laid eggs; and the process was repeated with each new generation of flies.

The flies Rose used had been living in the laboratory for 5 years before he started his experiment, and for those 5 years, the eggs used to start the next generation of flies had always been the ones produced on the 14th day after transfer. Females from this laboratory population lived on average about 33 days, and a fit fly was one that made lots of eggs at age 14 days. But that situation was about to change.

In 5 of his 10 populations, Rose changed nothing. New generations continued to be founded with eggs produced on the 14th day. These populations were the control populations. In the other 5 populations, Rose progressively increased the age at which the eggs used to start the next generation were collected; instead of gathering the eggs on the 14th day, Rose began collecting them on the 15th day, the 16th day, and so on. These populations were the ones experiencing Rose's artificial selection. All of a sudden, a fly wasn't very fit unless it could produce eggs at an older age; it didn't matter how many eggs a fly produced on Day 14, because none of those offspring were going to make it into the next generation.

After just 15 generations (15 transfers of eggs to new containers), Rose measured the life span of flies from the 10 populations and found that the flies from the populations selected for later reproduction lived an average of 20 percent longer than flies transferred every 14 days! By increasing the importance of events later in life, Rose had increased the strength of selection at later ages; as a result, the flies evolved to live longer.

Selection experiments always compare a control group of organisms with an experimental group. The experimental group experiences the artificial selection regimen, and the control group doesn't. In all other ways, both sets of organisms are treated exactly the same; they are kept in the same lab environment, handled by the same people, and so on. This technique eliminates doubt that any interesting results are caused by the artificial selection, not by some random factor (such as how hot the lab was that summer). As a further precaution, the same experiment with the same control treatments are performed many times. In experimental science, unless something happens several times, it doesn't really happen at all.

### Testing for metabolic trade-offs

With the two different groups of flies that he created in his life-span experiment, Rose tested the theory that metabolic trade-offs are involved in life-history evolution. Rose and his co-workers set out to look for evidence of these trade-offs in their selected flies. Here's what they found:

- The longer-lived flies had increased storage reserves of fats and carbohydrates compared with the control flies.

- The longer-lived flies had lower *fecundity* (lifetime reproductive potential) and devoted less energy to the production of eggs than the control flies did.

Bottom line: Rose's experiment produced evidence of the predicted life-history trade-offs: The longer-lived flies spent more energy on living and less on reproducing.

# The Trade-Off between Survival and Reproduction

From a fitness perspective, the best (that is, most "fit") reproductive scenario is one in which the organism begins to reproduce early and often, and all the offspring survive. But that's not the way it happens in nature, mainly because of the trade-offs between survival and reproduction. The resources required for survival are resources that aren't available for reproduction, and the resources necessary for reproduction aren't available for survival. So when and how often an organism reproduces, and how many offspring it produces, are the results of trade-offs that give that organism the best chance for sending its genes into the future.

You find various life histories in nature. Some organisms produce over and over again; some produce a few times throughout their lives; and some produce a single time and then die (perhaps the most extreme case of putting all your eggs in one basket!). Each species evolved in a way that allowed it to persist in its physical environment.

# It's good to reproduce often, except when it's not

Organisms that reproduce several times — humans, birds, and numerous other creatures — are called *iteroparous*. Organisms that reproduce only once are called *semelparous*. Semelparity is the most extreme example of the trade-off between survival and reproduction. The salmon is one example of a semelparous organism. After swimming back to the spawning ground where it was born, a salmon devotes every last ounce of its energy to reproduction; then it dies.

Two classes of environmental conditions lead to selection favoring the "reproduce once and die" genes:

✔ When getting a second chance to reproduce is pretty much impossible

✔ When some particular condition of the environment makes the reproductive payoff for reproducing once and then dying much greater than the payoff for reproducing over several years

The following sections examine these scenarios.

## One chance to make good

Reproducing just once before dying is the fittest thing to do when the odds of making it to the next reproductive season are very low.

After dodging all those bears and jumping all those waterfalls on the way to the spawning grounds, the friendly salmon is pretty lucky to have made it at all. If a gene appeared that made the salmon produce fewer offspring, with the expectation that it would swim back out to sea to feed again so it could reproduce more the following year, the fish would leave some descendents, but the odds of its leaving any more are fairly low. Instead, it might die as it tried to make it to the spawning grounds a second time. For this reason, this particular "produce less now so you can produce more later" gene isn't very likely to increase through time. In fact, from what scientists know about the risks facing the salmon, it makes sense that they don't see salmon with this strategy. Salmon are wonderful examples of the "reproduce once and die" strategy, but they aren't the only ones.

Lots of different organisms have a pretty low chance of making it from one reproductive season to the next. For many of these organisms, the culprit isn't anything nearly as picturesque as cascading waterfalls and hungry predators; it's the rather mundane reality of living in a seasonal world.

Imagine some tiny plant in the desert that's managed to germinate and grow during the wettest part of the year. If that plant is to have any chance of leaving descendents, it has to do it fast, while it still has enough water to survive. Because the adult plants can't make it to the hottest part of the year, they devote all their energy to reproduction, producing seeds that can withstand drought and soaring temperatures and that will germinate during the next rainy season.

### Big payoff for a one-shot deal

Most semelparous organisms — those annual desert plants, for example — are relatively short lived. They gather the resources they need to reproduce and get on with it. Other very long-lived semelparous species could reproduce earlier but don't; they just hang out, continuing to grow and acquire resources, all the while taking the chance that they'll die before they get around to reproducing. You can see the potential down side of this strategy. So what's the advantage?

Although the details vary from case to case, the bottom line is that peculiarities in the ecology of a particular species result in a disproportionate benefit to having a big bang in reproduction that outweigh the risks. The risk is big, but the payoff is also big, big enough that the genes for hanging back and stocking up are favored.

One of the best-known examples of this pattern is the blue agave *(Agave tequilana)*, the plant from which we distill tequila. The blue agave has a very high rate of year-to-year survival yet exhibits a life-history pattern of semelparous reproduction. (Although most agaves behave in a similar fashion, iteroparous agaves exist as well, but they're rare.)

When the blue agave finally decides to reproduce, it reallocates all its energy to reproduction and then dies. The selective pressures on this agave are different in nature from the pressures on the small annual desert plants growing all around. The little plants have to reproduce once because they can't survive the heat of the summer. The agave, on the other hand, can survive just fine in the summer, but it still has a greater chance of getting its genes into the next generation if it devotes all its energy to reproducing just one time.

The reason has to do with a peculiarity of agave reproduction. The agave needs something that's in short supply: *pollinators,* which are animals that move pollen to and from other agaves so that the plant can make seeds. When an agave reproduces, it makes a flower stalk. These stalks can be huge, reaching 25 feet in height. The agave uses every last bit of energy it has to make the largest, most visible floral display possible — a good plan, given that pollinators preferentially go to the most visible spike. If your spike is only half as big as another, you get fewer than half as many pollinators.

An agave with a gene that produced the trait of making a smaller spike and surviving until next year to make another smaller spike eventually would be eliminated from the population. Although that particular agave would keep surviving, it wouldn't make as many seeds because it wouldn't be as attractive to the animals that pollinate agaves.

## Early vs. later reproduction: Why wait?

An important component of an organism's life history is the way that the life span is divided between the pre-reproductive and the reproductive periods — or, in more common terms, between the time spent as a juvenile and the time spent as an adult. When scientists look at nature, they see a huge variation in the ages at which different organisms mature and become reproductive. Some organisms reproduce at very young ages; others wait until they are much older. Once again, organisms can do the same thing in many ways — in this case, producing offspring.

As I mention earlier in this chapter, you may think that the best strategy for promulgating genes would be reproducing early, making a lot of offspring, and doing all that for a very long time. But studies of the natural environment show that it's not possible to do everything well.

When it comes to reproduction and getting your genes into the next generation, it's not just how soon you can make the first offspring, but the total offspring you can produce throughout your lifetime. An organism that waits a little bit longer to get started but makes many more offspring than an organism that started earlier will have a greater effect on the genetic makeup of the next generation — that is, a higher proportion of organisms in the next generation will carry its genes.

The following list outlines the advantages and disadvantages of both strategies:

✔ **Reproducing early:** The benefits of early reproduction seem obvious. Genes involved in early reproduction will make it into the next generation that much faster. All other things being equal, earlier reduction would always be better than later reproduction. But early reproduction creates problems, the main one being that producing early saps the energy an organism needs for survival and growth.

✔ **Reproducing later:** An organism that delays reproduction has more resources to devote to survival than an organism that reproduces early. In addition, large organisms are often much better at making offspring than smaller organisms; they can produce more or healthier offspring. So it seems that some types of organisms get a real payoff from devoting energy to growth early in life and waiting until it's more efficient to begin reproducing.

In comparing the two strategies, it certainly appears that the better reproductive strategy is to wait until the organism is large enough to make many more offspring. In turn, these offspring should overwhelm the descendents of the individuals that reproduce early, which are wasting precious energy to maximize their reproductive output. But one pesky problem occurs: During the time that an organism is devoting energy to growth rather than reproduction, it could die, leaving no descendents. So which strategy is better: early or later reproduction? The answer is that it depends.

Scientists' understanding of evolution by natural selection suggests that as the risk of mortality increases, organisms evolve to reproduce earlier. Genes connected with early reproduction are favored and are more likely to make it to the next generation. Organisms with genes that result in later reproduction have a good chance of dying before they reproduce; as a result, these genes are not favored.

## Proving the point with guppies

David Resnick and co-workers set up an experiment in a natural environment to examine how altering the likelihood of surviving at different life stages could lead to evolutionary change in life history parameters, such as when to start reproducing. Resnick conducted his experiments with naturally occurring guppies in a series of streams on the island of Trinidad. The goal of the experiment was to see whether changing the relative importance of mortality for juveniles and adults would lead to changes in the guppies' life history. Specifically, Resnick wanted to see whether guppies' reproductive patterns, such as age at maturity, would change if their risk of death changed.

### The setup

The basic structure of the experiment was to figure out some way in nature to tinker with the survival chances of the fish. Resnick made use of a few things he knew about guppies in Trinidadian streams:

✔ **The major cause of mortality for guppies is getting eaten by two other types of fish.** For simplicity's sake, I'll call these species *the big predator* and *the little predator*. Although both species eat guppies, they prefer different sizes: The little predator likes to eat little guppies, and the big predator likes to eat big guppies.

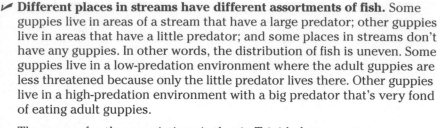

✔ **Different places in streams have different assortments of fish.** Some guppies live in areas of a stream that have a large predator; other guppies live in areas that have a little predator; and some places in streams don't have any guppies. In other words, the distribution of fish is uneven. Some guppies live in a low-predation environment where the adult guppies are less threatened because only the little predator lives there. Other guppies live in a high-predation environment with a big predator that's very fond of eating adult guppies.

The reason for these variations is that in Trinidad, many streams run down hills steep enough to produce waterfalls. Because fish can't get up waterfalls easily, the waterfalls effectively keep them from moving from one place to the next.

✔ **The guppies in the different habitats had different characteristics.** Resnick noticed that guppies living in the low-predation environments have different life-history characteristics from guppies living in high-predation environments:

- In the low-predation environment, guppies reproduced at a later age, when they were larger. They were able to invest energy in growth and survival that ultimately delayed reproduction — a successful strategy in the absence of a big predator. When these guppies made it to adult size, they had a reduced chance of being eaten.

- In the high-predation environments, the guppies reproduced earlier, when they were smaller. This strategy could be the result of predation pressure from the large fish. Delaying reproduction to a later date, when you're a bigger fish, isn't advantageous if you have a good likelihood of being eaten before you make any offspring. As the chance of being eaten increases, so does the selective pressure for early reproduction.

What Resnick saw was consistent with scientists' understanding of life-history evolution. As the chance of not making it to an older age increases, so does the advantage of reproducing earlier. Even better, not only were the results consistent with hypotheses about the evolution of reproductive life history, but this natural system was ready-made to test these ideas in nature.

### The experiment

Resnick found places in the streams that had little predators but neither big predators nor guppies. He proceeded to move guppies that had been coexisting with big predators into this new environment, where they had to contend with only a little predator. (To be sure that other fish wouldn't swim into his experimental areas, Resnick chose study locations in parts of the stream separated by waterfalls.)

# Natural settings vs. labs

The advantage of conducting artificial selection experiments in the laboratory is that it's possible to control all the variables of the experiment. The experimenters can change just one thing, hold everything else constant, and then see what the result of changing that one thing was. The goal of any experiment is to change as few things as possible so that when it's time to analyze the results, a lot of confusing questions won't be asked, such as exactly what caused those results. Another advantage of the laboratory is that it is easy to repeat the experiment to make sure of getting the same result a second time. The disadvantage, of course, is that laboratory experiments are always open to criticism that whatever happened in the laboratory doesn't really happen in nature. (For an example of a laboratory experiment, refer to "Methuselah flies: The evolution of life span in the laboratory" earlier in this chapter.)

Experiments conducted in nature overcome this problem. Any effects that are measured are obviously the result of natural processes. For this reason, they are important for developing scientists' understanding of how evolution functions in the wild. It's exciting to see that over a relatively short period, natural selection can lead to changes in populations. But this advantage comes at a cost: Changing just one thing is rarely possible. Controlling for all the variables is impossible, so natural experiments will always be open to criticism that the results were caused not by whatever the experimenter was manipulating but by something else.

Which type of experiment is better? Actually, neither. Researchers need laboratory experiments in which they can alter just a few things at a time, but they also need experiments in the field, where they can make some manipulations, let nature take its course, and see what happens.

After moving the fish, he waited to see how the guppies would evolve. If the pattern of early reproduction he observed among guppies in the presence of the big predator was due to increased predation pressure, moving the guppies to the different environment might result in the evolution of later reproduction.

After 4 years in one experiment and 11 years in another, here's what he found: The populations that were moved from a high-predation environment evolved their life-history characteristics in the expected direction. That is, the fish devoted more energy early in life to growth and less to reproduction. As a result, they were larger when they first reproduced.

### Determining whether the changes were heritable

To determine whether the changes were heritable — that is, genetic and therefore capable of being passed from one generation to the next — and not the result of some environmental factor, such as different food or different water chemicals in the various streams, Resnick took back to his laboratory some guppies from the old population that was living with the large predatory fish and some guppies from the experimental population that had been living with just the little predators. In the lab, he grew the guppies separately for several generations in identical fish tanks, to eliminate any effects due to the environment rather than the fishes' genes.

What Resnick found was that the fish really were genetically different. The fish from the low-predation experiment really *had* become different genetically from the original population. They spent more energy growing and reproduced later than the original population. The fish had indeed evolved.

### One fish, two fish, small fish, adieu fish: The evolution of overfishing

Understanding life-history evolution can help researchers understand some of the changes in major fisheries, where commercial fishing has made humans the major predator.

Different kinds of fish grow to different sizes, based on what's most advantageous in their environment: Tuna get really big; trout so big. A fish's life history involves growing for a while and then spending energy on reproduction. Natural selection favors variants that get the details right: Reproduce at too early or too late an age, and you won't be as successful as the fish that reproduces at just the right time. So the basic strategy is to grow, grow, grow until the point when it's better to start reproducing rather than growing some more. Bottom line: Fish in nature grow to some particular size because that's a good size to be as far as getting your genes into the next population.

Now along comes the fisherman, who sets out some fishing nets to catch fish. These nets aren't designed to catch just any fish, but bigger fish, and they're so effective that that they often cause a noticeable decline in the numbers of fish in the ocean. The fish numbers become so depleted that the fisheries stop or reduce their operations, with the intent of giving the fish stocks a chance to recover. Sounds all well and good, but guess what? The stocks don't recover as quickly as expected. Even after a lull, the big fish that the nets catch just aren't there.

Why not? What's happened is that a new selective force has been added to the environment. All of a sudden, being a fish of the size that gets caught in those nets is bad. What used to be a good size for getting your genes into the next-generation is now a good size for getting your genes grilled and covered with a nice lemon dill sauce.

Scientists and conservationists hypothesize that, in the past, before commercial fishing, genes for reaching adulthood at a smaller size hadn't been favored. To best survive a life in the ocean, fish evolved to grow to a sufficient size and then reproduce. Commercial fishing add a new selective pressure that makes it better to be a smaller fish. Their nets don't just physically remove the big fish from the population, but they also remove the genetic variants that result in larger fish, leaving only those that don't grow so big. Although being smaller may not be great as far as life in the ocean is concerned, it's the best thing going. Fish that don't grow big enough to get caught in the nets will pass more genes to the next generation. The result is a population of fish that just doesn't get so big. They reach adulthood at a smaller size and start devoting energy to reproduction. Just because we stop fishing doesn't mean that all the little fish that are left will grow up to be bigger fish; they've already grown up. And this is an explanation for why fisheries stocks don't rebound when we reduce the fishing pressure.

Evidence from natural fisheries suggests that the fish humans harvest have indeed shifted to earlier reproduction and smaller adult size. Northern cod populations, for example, exhibited a decline in the size at which females became reproductive before the collapse of the fishery.

To test the hypothesis that fish will evolve to be smaller if mortality at large sizes is increased, Matthew Walsh and coworkers conducted an experiment using laboratory populations of Atlantic silverside. They set up three different treatments and harvested fish differently in each one. From the control tanks they took a random sample of fish, from the other tanks they took large individuals — a protocol that mimics fishing. (To find out about the third treatment, read the sidebar "Fishin' for a small one.")

All the populations started out exactly the same — a bunch of wild fish in tanks. Over five generations, the fish populations responded to the new selective pressure. The results were exactly what the researchers had predicted. When large fish are more likely to be harvested from a population, the advantage of alleles that cause fish to grow large decreases. These alleles are less likely to make it into the next generation while alleles that result in smaller size now have higher fitness. These changes will cause the fish to be smaller, and that's exactly what happened. Selection favoring small fish over large ones resulted in fish that just don't grow as big. These changes included differences in metabolic efficiency, foraging, and even the number of vertebrae.

# Fishin' for a small one

Walsh and his coworkers also had a treatment from which they removed the smallest fish, which is not something any fisherman—commercial or otherwise—ever does when actually fishing for food, but it's a nice experiment anyway because the prediction is that removing the smaller individuals would cause the fish to spend less of their lives at the smaller size. Alleles that caused the fish to grow really fast would be favored because they'd be less likely to be removed from the tank by the experimenters. It's bad to be a small fish, and the only way to not be a small fish is to grow. The result of this treatment was bigger fish.

Figure 10-1 shows the result of Walsh's experiment. The fish on the right came from the population from which the largest fish were harvested. In the center are fish from the control tank. On the left are fish that came from the population from which small fish were harvested (see the sidebar "Fishin' for a small one for details on this part of the experiment).

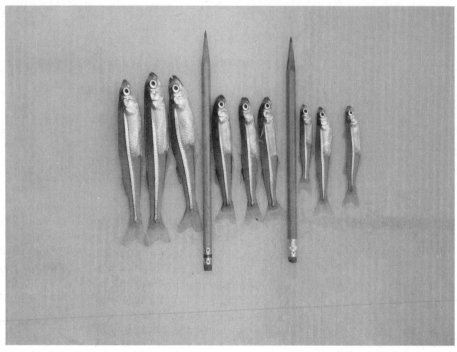

**Figure 10-1:**
The results
of Walsh's
experiments.

*Photograph taken by Stephan Munch*

The take-home message is two-fold:

✔ Over-fishing is bad for all the reasons we used to think it was, plus it's bad because the study of the evolution of fish suggests that fish stocks won't bounce back to include a large individuals as fast as we'd like them to if we stop fishing. We used to think that if we stopped fishing, the little fish we hadn't caught would grow up to be big fish. Now we have the added worry that we've selected for fish that don't grow up to be big.

✔ The remaining fish may not be as well adapted to their environment as they would have been had they not had to respond to fishing pressure. Fish that reach maturity at the smaller size are favored when there are lots of nets, but that may be the only thing good about a smaller size. The smaller size wasn't favored before fishing, and it may just not be a very good size with respect to doing all the things fish have to do to survive in the ocean.

# The trade-off between size and number of offspring

Just as a huge variation exists in life span and reproductive timing among organisms, a huge variation exists in fecundity and in *clutch size,* the number of offspring an organism produces at any one time. Trade-offs come into play here as well.

For any given amount of resources devoted to reproduction, the pie can be divided in many ways. (For fun, call this division *resource partitioning* among offspring.) It comes down to the question of whether a species produces lots of little offspring or fewer bigger ones. Human clutch size, for example, tends to be one, despite rare cases of twins or triplets, and total lifetime fecundity rarely exceeds ten. But a sturgeon — the kind of fish from which we get caviar — can easily make 60,000 eggs at a time. That figure makes humans look pretty pathetic in comparison until you consider that the human pattern seems to be successful; an awful lot of humans are around, and our numbers just keep going up.

What explains the huge differences in fecundity, and why haven't sturgeon taken over the world? The number of offspring produced is important, but so is the probability that these offspring survive to reproduce themselves. Any gene that results in the production of a huge number of offspring, but none of which ended up making any offspring of its own, wouldn't last very long. The gene's frequency would increase for an instant, but by the time of the grand-child generation, it would be gone.

As always, organisms go about producing fit offspring in many ways, favoring different strategies in different environments. Specific conditions favor different clutch sizes. Parental care — or the lack thereof — also has an impact, as the following sections explain.

## Without parental care

Not all organisms have a reproductive strategy that includes parental care. Sturgeons, for example, don't care for their young. After they release their eggs, they hit the road, leaving the eggs to fend for themselves. For these organisms, the question becomes how best to use the available resources for egg production to maximize the number of successful offspring. The organism can make a few large eggs or many small eggs. Whether big eggs or little eggs are best (or the most *fit,* in evolution-speak) depends on the environment the eggs will face.

At minimum, the egg needs to have sufficient resources to develop into a juvenile. Although an individual offspring's survival is increased if more than the minimum resources are provided, the parents' fitness depends on producing the maximum number of surviving offspring. Bottom line: How hostile or welcoming the environment is affects how much energy the parents invest in each individual offspring. Some environments result in a little extra provisioning of fewer eggs, but when the environment is hostile enough to make it unlikely for any particular offspring to reach maturity, the parents are better off spreading the risk among many eggs rather than investing a great deal of energy in each individual egg. The result? Smaller but more plentiful eggs.

This pattern occurs in sturgeons (60,000 eggs at a time, remember?) and in the plants commonly referred to as weeds, which tend to make large numbers of very small seeds that disperse through the environment. Where the seeds end up — whether in your freshly turned garden or in the middle of the highway — is anyone's guess. The point is they're scattered in the wind and have a low probability of survival.

Oak trees employ a different strategy. Although an individual tree makes lots of seeds, these seeds are quite large by plant standards. An oak tree doesn't make the maximum number of very small seeds; instead, it prepares the seeds — acorns — with a larger number of maternal resources. Acorns don't disperse very far from their mother plant. Instead, they tend to fall on the forest floor under a tree's canopy, and they need enough energy to sprout. With a little luck, they'll have a chance of getting big enough to gather enough light to survive. If the oak tree made just very small seeds, none would survive, because they would never have the energy to get big enough in the dark canopy of the oak forest. Thus, the environment facing the offspring of the oak tree favors genes for larger seeds, whereas the environment facing the weed seeds favors small seeds.

## With parental care

In the cases where natural selection favors organisms that provide parental care, it doesn't make any sense to produce more offspring than the parents can care for at any one time, because the parents continue to provide resources to the offspring after birth or hatching.

A fair number of studies have been conducted on why birds lay the number of eggs they do. The naturally occurring variation in the number of eggs produced gives scientists a sense of the optimal number for any particular species.

Ideally, the birds would produce exactly the number of offspring that they could feed well enough so that the offspring would survive and reproduce, passing the parents' genes into the future. What researchers find when they look at natural patterns, however, is a point in the number of eggs produced at which laying more eggs results in fewer surviving baby birds. The number of eggs that birds lay seems to cluster around this ideal number.

For an interesting case of how parental care can limit clutch size, think back to the emperor penguin. Penguin parents care for their single egg in a unique way: Dad balances the egg on his feet to keep it off the ice while Mom goes off to catch fish. Maybe the daddy penguin could balance a bigger egg or a smaller one, but it's not clear how he could balance two at the same time. Because no nest-building materials are available in Antarctica, an emperor penguin can't have more than one egg at a time.

# Why Age?

Organisms often seem to undergo a gradual breakdown before death — a process referred to as *aging.* Because aging, like death, seems to be a bad idea, you may think that natural selection would favor individuals that age less. Unfortunately (for humans), it doesn't.

The process of aging turns out to be consistent with scientists' understanding of how evolution works. You may think natural selection would eliminate genes that make people age, but natural selection won't eliminate certain classes of genes that cause aging, either because the detrimental effects of these genes don't show up until later or because the gene that causes aging offers some benefit earlier:

- ✔ **Bad genes that act later in life:** Selection acts less strongly on traits that are expressed late in life. Natural selection certainly won't favor any genes that cause the aches and pains of old age, but it can't select against them either. Long before the traits appear and can be noticed by selection, the genes that control them have already been passed to the next generation

- ✔ **Genes that are bad later in life but good when you're younger:** Natural selection favors genes that have beneficial effects when the organism is younger, even if these same genes are responsible for old age. The reason? Selection is stronger earlier in life. It doesn't matter what the genes do when you're old, because by then, you've already passed them to your children.

The two classes of genes are not mutually exclusive, and evidence exists that both types of genes are involved in the process of aging. Yippee.

# Conducting a thought experiment

If you're finding it a bit tricky to keep track of all these hypothetical classes of genes, try considering more specific examples. Conduct a thought experiment: Imagine a specific gene, and then try to figure out whether natural selection will favor it, select against it, or remain neutral. If you want to try to understand aging, for example, imagine the fate of a series of genes that does exactly the same thing, but at different times in the organism's life. In this way, you can get a better handle on how the frequencies of genes that act at different ages change in response to natural selection. Start with an easy one — say a gene the causes spontaneous combustion — and let your mind play with the ideas from there:

✔ Imagine a gene that causes people to spontaneously combust at age 10. What happens to people with this gene? It's safe to say that they won't be making a lot of offspring, so extremely strong selection will be against this gene. Because this gene causes death before reproductive age, it will be eliminated from the population as fast as new mutations make it appear. In other words, the genetic lifeguard says, "You're out of the gene pool!"

✔ How about a gene that makes people spontaneously combust at age 150? In this case, natural selection will never see the trait. None of us makes it to 150, so no one would ever go up in flames. This gene is completely neutral, and if it appeared in one of your children, you'd never even know it. Whether or not it increases in frequency depends only on whether your child has more or fewer children. Because the "combust at 150" gene has no effect on fitness, it's selectively neutral — just along for the ride.

✔ How about spontaneous combustion at 60? This, you'd notice. Would it make you less fit? Not really. By the time you're 60, you've pretty much finished passing on your genes. This gene wouldn't help you reproduce more, but it wouldn't hurt you, either — unless you happen to be one of the occasional people who burst into flames. Hence, the "combust at 60" mutation is also selectively neutral.

✔ How about a gene that made you spontaneously combust at age 60 but *also* made you much more likely to have lots of children? This gene increases your fitness because it increases your contribution to the next generation. Natural selection causes this gene to increase in frequency.

These last two categories of genes are the kinds of genes that would be responsible for the phenomena of aging. The negative effects of these genes occur only late in life, after they've already been passed to the next generation. In the last example, selection will cause the gene to increase in the population because it's advantageous early, even though it's really bad later.

Now step back from the example of spontaneous combustion, and think about some of the more real examples of things associated with aging. Nothing about your eyes failing or your knees getting creaky is at odds with the mechanism of evolution by natural selection. By the time those phenomena start to occur, you're probably finished reproducing. There may be better ways to build an eye or a knee, but natural selection won't favor those methods unless they have value earlier in life.

One last important note: Zero evidence exists that spontaneous human combustion really happens. It's in the same category as ESP and Roswell aliens. But just coincidentally, most of the people who are said to have spontaneously combusted were older than 60.

# Chapter 11

# Units of Selection and the Evolution of Social Behavior

· · · · · · · · · · · · · · · · · · · · · · · · · · · · · · · · · · · · · · · · · · · · · ·

· · · · · · · · · · · · · · · · · · · · · · · · · · · · · · · · · · · · · · · · · · · · · ·

*T*hroughout this book, I emphasize the importance of evolution acting on individuals. Why? Because the individual is far and away the most important unit on which selection acts. But natural selection doesn't act *only* on individuals. Sometimes selection acts at other levels. This chapter explores levels of selection: the power of natural selection to act on genes and groups.

Understanding the levels of selection is important in trying to fathom some of the behavior you see in nature. Some organisms, for example, don't reproduce. Others delay reproduction. Some behave altruistically toward others (giving up their own stores of food, for example, to prevent another organism from starving). In these systems, something besides individual selection is probably going on.

## Inclusive Fitness and Kin Selection

Evolutionary fitness is a measure of how good an organism is at getting its genes into future generations. One way for an organism to get its genes into future generations is to make lots of offspring, which go out and make lots of offspring themselves, and so on and so forth. But another way for an organism to get its genes into the next generation is to help its relatives get *their* genes into the next generation — which is the main idea behind inclusive fitness and kin selection.

# *Your fitness + your relatives' fitness = inclusive fitness*

*Inclusive fitness* is simply the sum of an individual's fitness *plus* the additional benefits accrued through increasing the fitness of related individuals. In English: You're more fit not only if you reproduce, but also if relatives who share your genes reproduce, too.

Here's a simple example of inclusive fitness in action: Helping your identical twin sister have a baby that she would not have been able to have without your help is just as good a way to pass on your genes to the next generation as having a baby of your own. Most humans aren't one half of an identical pair, of course. To understand how inclusive fitness works in more common situations, you need to understand a concept called *degree of relatedness,* which simply means how many of your genes you share with others.

For sexually reproducing organisms (such as humans, antelope, or titmice), the degree of relatedness is as follows:

- ✔ **Between parents and offspring:** One half of an individual's genes come from its mother and half come from its father. The degree of relatedness between parent (mother or father) and offspring is one half (or 0.5).

- ✔ **To full siblings (individuals that have the same mother and father):** Full siblings, on average, have half their genes in common. The degree of relatedness is 0.5.

- ✔ **To half siblings (individuals that have either the same mother or the same father):** Half siblings, on average, have one quarter of their genes in common. The degree of relatedness is 0.25.

- ✔ **To your siblings' children:** The degree of relatedness between you and any of your nieces and nephews is on average one quarter (0.25), because you share one half of your genes with your sibling, and that sibling shares half of his or her genes with his or her offspring.

So how many offspring do you have to help your siblings make to equal one of your own offspring? If you're an identical twin (that is, you and your twin have a degree of relatedness of 1.0 because you have exactly the same genes), helping her (or him) make an additional baby is just as good as making one of your own as far as your fitness is concerned. But if you're helping a non-identical full sibling reproduce at the expense of your own reproduction, to come out even you'd have to help produce at least *two* nieces or nephews for every one son or daughter you could have created yourself.

## *Not reproducing to help your family: Kin selection*

Nearly everything about evolution by natural selection involves some sort of cost-benefit analysis. Is the cost of the mutation offset by its benefit? Does having an elaborate tail help you more than hurt you? Natural selection favors the trait that confers the most benefit from a fitness perspective, meaning that traits that enhance your ability to get your genes into the next generation increase in frequency in subsequent generations.

The situation is no different for reproduction. If, for some reason, you were unable to produce any offspring of your own, helping produce even one extra niece or nephew would still enable you to get some of your genes into the next generation. Or if by forgoing producing a single offspring of your own, you were able to help produce three nieces or nephews who would not otherwise have existed, you come out ahead, evolutionarily speaking.

Because the cost (the single offspring that you did not produce who would have shared half of your genes) is less than the benefit (the three nieces or nephews who each share one quarter of your genes), natural selection will favor helping your sibling reproduce. For this reason, genes that are responsible for behaviors that help relatives reproduce can increase in frequency even if they decrease individual fitness as long as they increase inclusive fitness. This type of selective force is called *kin selection*.

The concept of kin selection explains many of the altruistic behaviors that Charles Darwin found confusing. For Darwin, who lacked modern knowledge of genetics, it wasn't as obvious how helping a related individual could be advantageous to the helper.

CASE STUDY

## It's a salamander-eat-salamander world

David Pfennig raised tiger salamander larvae in groups with and without siblings. Tiger salamanders are cannibalistic. For a tiger salamander, eating one of your own kind is good if it increases your chance of getting your genes into the next generation — but not as good if the one you eat has your genes.

In the study, Pfennig raised his salamanders in groups of full siblings, half siblings, and unre-

lated individuals. The results: Full siblings were least likely to cannibalize their neighbors, and unrelated individuals were most likely. Half-sibling groups were in between.

The result makes sense from the standpoint of inclusive fitness and kin selection. Genes that allow a salamander to tell kin from stranger (and to avoid eating the kin) appear to have been selectively favored.

# Levels of Selection

In individual selection, some individuals survive to reproduce, and some don't. If the differences in survival and reproduction are the result of particular genes, these genes will increase in frequency in the next generation. *Group selection* theory adds another evolutionary layer to this tidy little setup. The idea is that in addition to selection at the individual level, selection can also occur at the group level. (You can even think of kin selection as an example of group selection where the group is the family.)

Evolutionary scientists have also become aware that selection can act at the level of the gene, beyond the forces of individual or group selection. Essentially, this theory recognizes that occasionally, certain genes increase in frequency simply because they've got a better-than-average chance of getting into the gamete during reproduction, regardless of the negative effect they may have on the offspring.

# Group selection

According to group selection theory, some genes have an effect at the level of the group. Genes that act at the group level affect survival and reproduction between groups in such a way that some groups leave descendants and other groups don't. Think about a gene for getting along nicely with other people, for example. If selection acts at the group level, a get-along-with-others gene may increase in frequency if groups of people who have this gene are more successful than groups of people who don't have this gene.

As you wrap your brain around group selection, keep these points in mind:

- ✔ Group selection, when it occurs, acts *in addition to* selection at the level of the individual. Selection still occurs at the level of the individual.

- ✔ Selection at the level of the individual could act in the same direction as selection at the level of the group or in a different direction. For evolutionary biologists, the most intriguing cases are those where different levels of selection conflict and as a result are of considerable interest.

- ✔ Group selection can be important when the population structure of the species involves interacting groups.

### Flour beetles: A group selection example

Michael Wade set up a laboratory experiment to investigate group selection with the goal of showing that it occurs. In his experiment, Wade used flour beetles — little insects that are happy to grow and reproduce in a vial of flour. They eat flour, but they're also cannibalistic: Adult beetles eat larvae and eggs, and the larvae eat eggs.

Wade set up four experimental treatments, each consisting of 48 populations of beetles. For each population, he allowed 16 individual beetles to grow and reproduce for 37 days. Then he took 16 individuals from each of these populations, put them in new vials of flour, and allowed *them* to grow for 37 days. He continued in this way until he had the number of populations he needed. What differed between the populations in the various treatments was the factor that Wade selected for. For his experiment, he decided that population size — a group rather than an individual characteristic — would be the favored characteristic:

✔ **Treatment 1 (the control treatment).** In this treatment, individual selection was the key. For the control group, Wade randomly picked 16 beetles from a population to start the next flask. As a result, individual beetles that were more successful at reproducing were more likely to have offspring — and, hence, their genes — in the next generation. (If you make twice as many offspring as the next beetle, you have twice the chance of getting some children into the next generation.)

Individual selection is always happening, so scientists can study group selection only by keeping track of individual selection at the same time and then comparing the results from the group selection treatments to the individual selection control treatment. The control group is the one in which selection acts at the individual level.

✔ **Treatment 2 (a group selection treatment selected for large population size).** Wade created his new populations by using beetles from only the biggest populations.

✔ **Treatment 3 (a group selection treatment selected for a low population size).** Wade used only individuals from the smallest populations to found new populations.

Treatments 2 and 3 represent group selection, because not all beetles in all vials contribute to the next generation — only those that meet the selection criteria (in the experiment, either large or small group size). You could be a really fit beetle in your particular vial (that is, a disproportionate number of the offspring are yours), but if you exist in a population that isn't large (or small) enough to be used, none of your genes get into the next generation.

Here's what Wade found: The original beetle stock used in this experiment generated population sizes of about 200 beetles after 37 days. But after nine generations, population size decreased in all his treatments. Population dropped in the control treatment (individual selection); it decreased more in the group selection treatment for small populations and less in the group selection treatment for large populations. The following sections explain why.

### In the control treatment (Treatment 1)

In this case, selection was acting at the level of the individual. Whatever heritable traits increased the chance of a beetle's having descendents in the next generation increased in frequency. It just so happened that as a result of this selection on individuals, population size decreased from about 200 beetles in the stock population to 50 beetles after 9 generations. Although that result may seem odd, keep in mind what these beetles like to eat: flour and baby beetles (those in the larval and egg stages). Essentially, this treatment selected for beetles that were more voracious cannibals.

### In Treatment 2 (group selection favoring large population size)

In this treatment, population size had decreased after nine generations, but not as much as in Treatment 1 (the individual selection control). Selection at the level of the group for larger population size had an effect in the opposite direction of selection at the level of the individual.

Selection at different levels can act in the same direction or in a different direction. If selection acts in the same direction, it compounds the effect; if it acts in a different direction, it mitigates the effect.

### In Treatment 3 (selection favoring groups with small populations size)

In this treatment, population size decreased even more than in the control population. Selection at the level of the individual and selection at the level of the group each had an effect on population size, and these effects were in the same direction. After 9 generations of selection, the population had decreased from 200 beetles to 19.

## Selecting for nicer chickens: Applications of group selection

Although group selection can seem awfully technical, it has practical applications. Take, for example, egg production. A great deal of effort has gone into producing chickens that lay lots of eggs. Historically, farmers have tackled the problem at the level of the individual chicken. The chicken breeder would found the next population with the descendants of the most prodigious egg-layers. The strategy was successful: Chickens can lay 100 eggs a year.

Here's a little bit of detail that you should know about chickens. They don't live isolated lives on the farm. They are raised in group pens, where interaction among chickens can be downright nasty — so much so that chickens often have their beaks removed to prevent them from pecking one another. If the best egg-layer lays many more eggs than the other chickens because she takes resources away from them, you can easily see why the other chickens in the pen won't produce as many eggs as the individual egg-laying champ. What the egg farmer wants, however, is to maximize the total number of eggs produced. Because chickens are raised in pens, the important number isn't eggs produced by any particular chicken, but eggs produced per pen.

William Muir and coworkers wanted to see whether it was possible to increase the number of eggs that the chickens laid as a group rather than merely increasing the number of eggs laid by individuals. So they expanded on the farmer's breeding practices, which favored the chickens that laid the most eggs, by introducing selection at the group level.

Muir selected for the largest number of eggs per pen. Chickens from pens that produced a large number of eggs were used to found the pens for the next generation. The best-producing pens of that generation were the source of individuals for the subsequent generation. After five generations, Muir was able to produce chickens that got along so nicely with the other chickens in the pen that removing the beaks was no longer necessary. Just as significantly, egg production increased by 100 percent, to 200 eggs per year per chicken.

## Selection at the level of the gene

Selection can also work at the gene level. When a *diploid individual* (one that has two copies of its genome) produces *haploid gametes* (sperm cells or eggs, which have only one copy of the genome), each parental gene has a 50-50 chance of ending up in the gamete population — usually. Sometimes, though, for reasons that are poorly understood, one gene is better than another at making it into the gametes. This phenomenon is selection at the gene level, or *meiotic drive*.

Meiotic drive occurs when a particular gene has better than a 50 percent chance of making it into the gamete pool and from there to the offspring. Although on its face, selection at the gene level seems to be at odds with how you expect evolution to work, by taking a closer look at meiotic drive, you can see that the same evolutionary forces are at play.

### Increases in subsequent populations

In nearly all the other discussions in this book, the probability of a gene's ending up in the next generation depends on whether that particular gene increases the survival and/or reproduction of the individual. The gene that makes a cheetah run fast, for example, is more likely to make it into the next generation than the gene that makes a cheetah run slowly.

Meiotic drive is different. The frequency of a gene increases because it ends up in a disproportionate number of offspring — not because it necessarily increases the organism's fitness. Even a gene that has a negative effect on the survival and reproduction of the organism could increase in frequency if the degree to which it disables the individual is compensated for by its increased representation in the offspring. In other words, individuals carrying this gene may be less fit, but there are more of them than individuals carrying the gene left behind by meiotic drive.

Any gene that's better at ending up in the offspring will increase in frequency, even if ending up in more offspring is the only thing it's better at doing and even if it has fitness costs for the individual carrying it.

### When selection levels collide

If the gene that increases in frequency has a high fitness cost to the individual, the direction of selection can be different at different levels. Suppose that a gene that reduces the speed at which cheetahs can run is one that, through meiotic drive, has a better than 50 percent chance of ending up in the gamete population. Selection at the level of the individual will act to remove such a costly gene, because cheetahs with this gene will be slower and will produce fewer offspring.

Because the fitness of a particular gene can be at odds with the fitness of the individual that carries it, selection at the level of the individual acts to combat the deleterious driver gene. Take, for example, fruit flies.

A particular species of fruit fly has an X-linked driver gene. This driver gene causes males to have only female offspring. Because producing both male and female offspring is the most successful strategy for getting genes into the subsequent generations, female flies are at a disadvantage if they mate with the male flies carrying this driver gene.

Interestingly, the female fruit flies choose mates that have long eye stalks. The connection? The driver gene is associated with a gene for short eye stalk. The female preference for the long eye stalk evolved because the length of the eye stalk correlates to the quality of the mate. Mate with a short-eye-stalk fruit fly, and you end up with only daughters and decreased fitness. But avoid short-eye-stalk fellas, and you get both male and female offspring and increased fitness. (For a down-and-dirty explanation of sex selection, head to Chapter 12.)

Selection at the individual level operates to oppose selection at the level of the driver gene. Females with a preference for males with long eye stalks are less likely to mate with males carrying the driver gene. So even though the gene is in twice as many gametes (males containing the driver gene make only gametes that carry this gene), it doesn't have twice the chance of making it into the next generation because the male fly isn't as likely to be chosen as a mate.

# The Evolution of Altruistic Social Systems

As I explain throughout this book, evolution by natural selection means that nature selects or favors heritable traits (or characteristics) that increase an organism's fitness — how successful it is at getting its genes into future generations. Yet some organisms actually postpone reproduction or forgo it.

At first glance, such actions seem to be absolutely out of step with the core concepts of evolution until you realize that in some cases, such as those highlighted in the following sections, delaying or abandoning reproduction can actually make an organism *more* fit. Which brings me to the evolution of altruism and how doing good for others can be good for you.

## Cooperative breeding

Although not common (only about 300 out of approximately 10,000 bird species have cooperative breeding), cooperative breeding does happen. Birds with helpers, regardless of whether the helpers are related or not, typically raise more offspring — sometimes twice as many — than birds without helpers.

There are two kinds of cooperative breeding among birds:

- **Non-reproductive individuals helping other birds to reproduce:** In some cases, the helpers are related; in other cases, they're not. Helpers who are related benefit because doing so increases their inclusive fitness (refer to the earlier section "Your fitness + your relatives' fitness = inclusive fitness" for details on inclusive fitness). Unrelated helpers benefit because if male dies they have the inside track on widow: wife and home all packaged together.

- **Groups of reproductive individuals getting together and helping each other (often with some non-reproducing individuals, too):** An example of this type of system is found with the groove-billed ani, as explained in the section "Other cooperative breeding behaviors."

### Helpers at the nest: When the helpers are related

*Helpers at the nest* refers to a situation in which offspring don't leave the parental environment immediately to raise offspring of their own, but remain with the parents for some period of time and assist the parents in raising additional broods.

Waiting to reproduce may be beneficial for birds whose probability of reproductive success is low. Suppose, for example, that reproductive success is limited by the number of available territories in which pairs of birds can breed. A younger, less-experienced bird may not be able to get and hold any territory;

therefore, on his own, he won't have much opportunity for reproduction. If he stays at the parental nest and helps care for his younger siblings, however, he can boost his inclusive fitness.

The reason bird studies tend to focus on the behavior of males rather than females is because it's usually males that defend territories and females that disperse to find mates — you know, the male bird sits and sings and hopes a female flies by. When territories are scarce, extra males pile up around home. In addition, the fact that the females are on the move can also mean increased female mortality, which skews the sex ratio of males to females.

Sometimes, no evidence exists that the helpers at the nest increase the number of offspring the parents are able to raise. In these cases, selection favors helping for different reasons:

- **To gain experience:** A wealth of evidence shows that inexperienced pairs of birds are much less successful at raising offspring than are experienced birds. By sticking around to help, the young birds gain experience in parenting.

- **To inherit the territory:** In an environment where territories are both limited *and* needed to successfully raise offspring, staying in the parents' territory may result in a better chance of inheriting the territory when the parents die. (Consider this the suck-up principle.)

- **To wait out tough times:** Environments fluctuate, and sometimes there just isn't enough food or enough available mates. In this case, hanging around in a bad year may increase your chance of surviving to breed in a good one later. Got a 30-year-old child living at home? Then you're intimately familiar with this scenario.

The young bird isn't sticking around just to be helpful; he's looking out for himself. The fact that his parents or siblings may benefit from his presence is beside the point. For that matter, the parents aren't acting altruistically either. Even if having the young bird around doesn't increase their own number of offspring, the parents still benefit if, by delaying leaving home, the young bird can increase his own reproductive success.

### When the helper isn't related

In some species, the male bird that helps out the mother bird isn't one of her sons or related to her at all. How can helping a complete stranger make more offspring be selectively advantageous?

In these species, mortality can be quite high. If the breeding male dies while the unrelated male is hanging around being oh-so-helpful, who do you think the female will turn to as a new mate? You've got it: Mr. Helpful himself, who, by assisting her, has given himself the inside track on the job!

### Other cooperative breeding behaviors

The groove-billed ani is a kind of cuckoo that lives primarily in Central and South America. These birds live in small groups (between one and five pairs) within a single nest. They all help defend territory, provide food, sit on the eggs, and feed the young. But this isn't an avian utopia. Before they lay their own eggs, the females make room for their own eggs by kicking others' eggs out.

The ani engages in cooperative breeding, but it isn't an example of kin selection because the helpers are other reproducing birds; they don't forgo or delay their own reproduction to help the others. Even with the problems of this system (the danger of your eggs getting tossed by another female, for example), each reproducing pair produces more offspring than it otherwise would have.

# One good turn deserves another: Reciprocal altruism

An individual can increase its own fitness through *reciprocal altruism,* the situation in which altruistic acts are repaid. Forgot your lunch? No problem. I'll give you half of mine. But at some point in the future, I'll expect you to return the favor. Unlike kin selection (described in the earlier section), in which the good-deed-doer gets an immediate benefit, in reciprocal altruism, the benefit comes at some future time. In addition, reciprocal altruism can occur whether the individuals are genetically related or not.

### When it's more likely to occur

Natural selection favors traits for reciprocal altruism (doing a good deed today in the hope of being repaid at some point in the future) under certain conditions:

- ✔ The individual that benefits from the altruistic behavior will be around in the future to do a good deed in return.

- ✔ The benefit you accrue from doing a good deed is greater than the cost of doing the good deed in the first place.

- ✔ Individuals are able to identify cheaters — individuals who accept altruism but never pay anyone back.

Because these conditions occur rarely in nature, instances of reciprocal altruism are rare as well. Still, examples of reciprocal altruism exist in nature. One naturally altruistic creature is — believe it or not — the female vampire bat, as shown in an experiment performed by Gerald Wilkinson.

### Reciprocal altruism in action

Female vampire bats live in colonies made up of unrelated females and their offspring, and they feed exclusively on blood. At feeding time, these bats leave the roost, find a mammal (usually, a nice juicy cow), bite it without being detected, and lap up the blood. On a good night, they can consume almost one third of their body weight in blood before they head back to the roost — which is like a 150-pound person having a 50-pound dinner.

But if a bat can't find food (sometimes, no cows are to be had), it's in big trouble. If it goes more than two nights without eating, it starves. In fact, studies have shown that vampire bats fail to find food often enough that, left on their own, most would be dead within a year. Yet vampire bats tend to live a long time. Why? Vampire bats share the wealth.

Back at the roost, the bats who were successful share some of the blood they collected with the bats that are starving. Then, when they themselves are on death's doorstep, they get a sip of blood in return. This exchange among the bats meets the conditions necessary for altruistic reciprocity:

✔ **The colonies are long-term communities of the same individuals who have many opportunities to share blood as needed.** In one case, the same pair of bats lived together for more than 10 years. Familiar bats — those with which the donor has a longer association — are more likely to receive a donation of blood, because they're the ones most likely to be around to reciprocate in the future.

Gerald Wilkinson removed a sample of bats from a colony and held them without food until they reached the weight at which they would solicit food from other bats. Then he compared how the bats remaining in the colony responded to their returning (hungry) compatriots and how they responded to hungry bats from a different colony. He found that the full bats were likely to feed the hungry bats with which they had previous associations and were much less likely to feed the bats that were strangers.

✔ **The altruistic act is more beneficial to the recipient than harmful to the giver.** When a bat is close to starvation, a little bit of blood can buy a fair bit of time. But that same amount of blood isn't as big a loss to the donating bat that's flush from a successful feeding trip.

✔ **Vampire bats seem to have mechanisms in place to make sure that other bats can't cheat:**

- They're able to recognize and remember other individual bats.

- Before a hungry bat solicits food from one of its roostmates, it grooms that bat's belly. Given that full bats are noticeably more rotund than starving bats, the hungry bat probably has a good sense of who has food to share.

- The bat that's being asked for food can assess the condition of the bat that's asking. Roostmates that are close to starvation are much more likely to be fed than roostmates that are only a little hungry.

# Going to extremes: Eusociality

*Eusociality* is a social system characterized by reproductive specialization. One or some individuals in the colony bear all the offspring, and non-reproductive individuals assist in caring for the offspring. This system sounds nice and tidy, but evolutionary biologists had to work long and hard to figure out why eusociality works or what benefits it confers to the workers to make them want to hang around rather than flit off to start their own colonies.

The biggest question is why a non-reproductive worker class exists at all. These individuals aren't postponing reproduction until sometime in the future (as is the case for birds helping at the nest). They're forgoing reproduction entirely. Their genes will make it into future generations only to the extent that they are able to help related individuals reproduce and their offspring survive.

For some eusocial species, like ants, bees, and wasps (all *Hymenoptera* species), kin selection and inclusive fitness explain why a nonreproductive worker class can develop, and it all goes back to a reproductive system that makes individuals more related to their siblings than to their own offspring. For the other organisms, researchers suspect that other forces are at play.

## Kin selection in bees, ants, and wasps

Bees, wasps, and ants live in highly structured colonies, with each individual performing particular tasks. Each colony, for example, contains a single reproducing female and many non-reproducing females that gather food, feed the young, and defend the nest but don't lay eggs of their own. So how does helping the group help (that is, increase the fitness of) the individuals, especially those that forgo reproduction themselves?

You could say that all these insects are working toward the good of the colony (and every animated ant or bee film ever made offers this explanation), but you'd be wrong. Evolution rarely works at the group level; it is more likely to work at the individual level. Each ant, wasp, or bee is really working toward its own benefit — not the colony's.

In these species, non-reproductive females (the workers) increase their own fitness more if they help the queen reproduce than if they reproduce themselves. It's a numbers game. Of all the relationships (parent to offspring, sister to sister, sister to brother), females in these systems share more genes with their sisters than they do with either parent, their brothers, or their own offspring. More sisters mean more fitness. The following sections explain.

### Degree of relatedness in Hymenoptera species

In *Hymenoptera* reproduction, fertilized eggs become females; unfertilized eggs become males. Because they result from fertilized eggs, females are diploid. A female, being diploid, gets half of its genetic material from each of its parents, just as humans do. What's different is that the daughters receive exactly the same genes from the father, because the males, having arisen from unfertilized eggs, have only a single copy of the genetic material to pass on to their children.

The upshot of this strange sex-determination mechanism is that females in this reproductive system are three times as closely related to their sisters as they are to their brothers, and mothers are more closely related to their sisters (three quarters) than they are to their offspring (one half). In terms of kin selection, being more related to your sisters than to your own offspring means that a female can increase her fitness more efficiently by helping her mother make another reproductive daughter than to make a daughter herself.

And guess what? That's exactly what happens in many *Hymenoptera* species. That beehive, anthill, or wasp's nest in your backyard usually contains one reproducing female and many non-reproducing daughters that take care of the nest and the babies but don't reproduce themselves.

### Intrigue in the queen's court

The mere fact that non-reproductive workers aren't laying eggs on their own doesn't mean that they're not affecting the reproductive decisions of the hive. Consider bees as an example.

The queen bee controls which eggs get fertilized, using sperm she stored from her maiden mating flight. Because her sons and daughters each share half of her genes, she would just as soon make half of each. It's all the same to her. But regardless of the sex ratio of the eggs laid by the queen, the ratio of reproductive bees produced by a hive often has a three-to-one bias in favor of females. So if half of the Queen's offspring are males and half are females, how does the hive end up with many times more females than males?

The answer? Her daughters, the worker bees. For their own fitness (being three times more related to their sisters than to their brothers), they prefer that the queen make more daughters. And because the worker bees control the feeding of the larvae, they control how many fertilized eggs will become reproductive females and how many unfertilized eggs (males) will survive.

Genes that result in worker bees producing more sisters are selectively favored because the sisters are likely to have those genes. A gene that favors producing more brothers is not as likely to get passed to future generations, because the brother bee is not as likely to share that gene (or any particular gene) with the sister.

### Other eusocial organisms

Not all eusocial organisms have the same reproductive system that *Hymenoptera* species do. Termites, for example, live in colonies with a reproductive pair (one female and one male) and sterile female workers, but the workers aren't more closely related to their siblings than they are to their own offspring (were they to have any). For them and other such creatures, kin selection isn't what compels them to remain in the colony.

In trying to understand these other eusocial systems, biologists suspect that eusociality may have been the result of a dynamic between selection at the level of the individual (which favors the fitness of the individual and therefore tends to prevent the formation of such systems) and selection at the level of the group (which favors groups of organisms over individual ones).

If playing a minor role in a larger group makes an individual more fit than playing a solitary role, natural selection could favor genes for remaining in the group. Think of the dancers in a chorus line. They may never be the headliners that draw in the audiences, but the show can't go on without them — and they're dancing on Broadway.

So how do roles get assigned in such species? What mechanism dictates whether a budding young organism is the star or a background player? Genes. Specifically, genes for *plasticity* — those that allow individuals to develop into reproductive individuals or non-reproductive workers. If such genes led to increased efficiency in the division of labor within the colony, they would be selectively favored.

## One is the loneliest number: Multicellularity

The fossil record makes it clear that organisms started out as single-celled entities and developed into multicelled organisms as time passed. It's not hard to imagine the evolutionary pathway that led from one condition to the other:

1. A single-celled organism divided into two daughter cells.

2. Each daughter cell divided into two daughter cells.

3. Each of those daughter cells divided, and so on and so forth, until a whole bunch of single-celled organisms were going happily about their business.

4. One day, a mutation occurred, resulting in two daughter cells remaining attached instead of separating after cell division.

These two still-attached cells had a couple of advantages over individual cells — maybe they were harder to eat (because they were bigger), and maybe they had more access to food sources — so this particular mutation increased in frequency in the population.

This example seems fairly simple, but making a multicelled organism isn't as easy as assembling a collection of identical cells. In larger organisms, different cells have to become specialized to perform different functions. The cells in your own body, for example, are remarkably good at sharing resources, yet few of them do any of the reproducing; that task is left entirely to the tissues that produce eggs and sperm. Yet the specialization that makes complex functioning possible can also lead to conflicts between different cell lineages.

Cells can be selfish in a couple of ways:

✔ **By not performing their appropriate functions:** If all the cells want to be gonads, and none of them wants to be a thighbone, the community (the multicelled organism) breaks down. The only chance that one of your nose-hair cells has of getting its genes into the next generation is to play nice, do its job, and not suck up any more resources than necessary.

✔ **By trying to hog all the resources:** Every once in a while, a mutation in a cell can result in that cell's breaking ranks and going all out for the resources. Need a recognizable word for this phenomenon? Try *cancer*.

Part of the evolution of multicellularity — organisms growing from 4 cells to 8 cells to 16 cells to 32 to eventually billions of cells — seems to be the natural selection of traits that keep selfish cells in check. For the most part, things have gone swimmingly. Organisms have evolved from single-celled creatures to multicelled creatures of amazing complexity and variety that can, for example, write or read a book about evolution.

# Chapter 12

# Evolution and Sex

. . . . . . . . . . . . . . . . . . . . . . . . . . . . . . . . . . . . . . . . . . . . . . .

## In This Chapter

▶ Understanding why sex is poorly understood

▶ Balancing the costs and benefits of sex

▶ Fighting for (or preening for) a mate

▶ Recognizing potential conflicts between the sexes

. . . . . . . . . . . . . . . . . . . . . . . . . . . . . . . . . . . . . . . . . . . . . . .

*E*volutionary biologists spend a fair bit of time thinking about sex, and in this chapter, you find out what it is, exactly, that they're thinking about. The evolutionary biology of sex encompasses several areas that can be conveniently broken down into the following major topics:

✔ How does the added complication of mating affect the process of evolution in sexual species?

✔ Why have two sexes, and what's the optimum male-to female ratio?

✔ Why have sex at all?

Just as this chapter covers a range of topics, it also illustrates a range in scientists' understanding of these topics. The evolution of sex is a hot area of research, and knowledge is growing rapidly. In some cases, researchers are confident that they understand what they see in nature; in other cases, they're not. In this chapter, I tell you what scientists know and what they're not yet sure of. As always, I continue to introduce you to the data and experiments that shape current understanding.

## Sex Terms You Probably Thought You Knew

The term *sex* can mean several things. Ask a seventh-grader straight out of health class, and you're likely to get one definition (with a few sniggers thrown in). Ask a parent of a brood of children, and you're likely to get another. Often, these definitions equate sex with sexual reproduction. But ask an evolutionary biologist to define *sex* and *sexual reproduction,* and you get definitions that are beautifully precise, even if they do suck all the fun and innuendo out of the terms.

For evolutionary biologists, *sex* is not mating but the combining of genetic material from different individuals, and *reproduction* simply means making offspring; reproduction, in other words, doesn't have to include sex:

 ✔ **Sexual reproduction:** *Sexual reproduction* is the system in which two individuals mate and produce offspring whose genes are a combination of some of the genes from each parent. Sexually reproducing organisms produce offspring that have only half of each parent's genes.

 ✔ **Asexual reproduction:** *Asexual* organisms produce offspring without mating, and these offspring contain only the single parent's genes.

For some species (like humans), reproduction *always* relies on sex. Other species (like rotifers and a few kinds of lizards) never have sex at all and still manage to reproduce. Still other species can reproduce with *or* without sex, proving that sexual reproduction isn't necessarily an all-or-nothing proposition.

As mentioned in numerous places throughout this book, *fitness* is all about passing on your genes to the next generation. The more successful you are at that task, the more fit you are. All other things being equal, an organism that makes twice as many offspring is twice as fit. (Go to Chapter 2 for a more in-depth discussion of fitness from an evolutionary perspective.) But in reproduction, the quality of the offspring's reproductive success also impacts an individual's fitness. If two individuals each produce four offspring, but the four offspring of one individual die before getting a chance to reproduce, the two individuals aren't equally fit.

# Sexual Selection: The Art of Picking a Mate

For sexually reproducing species, it's not enough just to find food and dodge predators. You also have to get a mate, because the prime directive, evolutionarily speaking, is to pass on your genes. How you go about choosing the best candidate to help you do just that is the topic of sexual selection.

*Sexual selection* refers to choosing certain characteristics over others when looking for a mate, and it's a subcategory of natural selection — the process by which heritable traits that promote survival become more common.

Sexual selection can lead to the evolution of some traits that, on the face of it, don't make much sense. It may favor a trait that increases an individual's ability to find a mate but decreases its ability to survive, for example. Although this arrangement seems strange at first, it's perfectly reasonable if the trade-off is a good one — that is, if the fitness increase from increased mating is greater than the fitness loss from earlier death.

Sexual selection has two components: choice and competition. For the vast majority of cases, this means female choice and male-male competition. These are not mutually exclusive; you can see both in the same species and sometimes you even see the reverse (male choice and female-female competition). Also remember that sexual selection is found in lots of animal species from insects to mammals.

Some evolutionary biologists argue that sexual selection should be considered separately from natural selection, as the evolutionary changes produced are of a slightly different nature (flashy feathers as opposed to anything that's actually helpful for surviving, finding food, and so on.) But for our purposes here, I treat sexual selection as a subcategory of natural selection.

# The Peacock's Tail: Sexual Selection and Female Choice

One kind of sexual selection system is one in which females choose their partners based on some outward characteristic. In these cases, the females' choice results in the males having some increasingly elaborate trait, even when that trait doesn't offer an obvious advantage in the male's survival. At face value, it's the classic example of choosing style over substance. This phenomenon has been puzzling evolutionary biologists since Darwin, because although it's obvious that females choose showy males, it's not entirely clear why.

In these species, evolutionary biologists figure that something else must be at play — something that makes it advantageous for the male (with the showy trait that serves no other purpose) and the female (whose choice of mate means that her offspring will end up with a trait that may actually be a disadvantage) to get together.

To understand this system, consider peacocks. If you've had the opportunity to see a peacock, you've no doubt been impressed by the size and beauty of his tail. The huge fan of iridescent green and blue feathers, which he displays so proudly, is unquestionably one of the most magnificent displays in the bird family.

That said, from a survival point of view, the fancy tail doesn't appear to have a lot of advantages. In fact, the tail seems to increase the chance that the peacock will be caught and eaten by predators. Peacocks don't fly especially gracefully, and the tail certainly makes it hard to hide. In addition, although the tail can't possibly make finding food any easier, the energy required to produce such an elaborate tail means that the peacock needs to find more.

In short, a showy tail doesn't seem to have any obvious fitness advantages, but it turns out that there is one, and it's a big one: Peahens like it! In fact, peahens like flashy tails so much that they preferentially choose to mate with males that have showier tails. Therefore, males with showier tails are much more likely to pass on their genes to the next generation than are males with less-showy tales. So having a fancy tail does indeed make a peacock more fit.

But what about the peahen? How does mating with the showy male increase her fitness? After all, she's deliberately choosing to mate with a very showy but otherwise less capable male. Although she's passing on her genes to the next generation (good from a fitness perspective), she's combining them with some genes that appear to make it more likely that her offspring will get eaten or starve (bad from a fitness perspective).

Two major ideas explain the curious choices that the females seem to make (and shallowness isn't one of them): the runaway-selection hypothesis and the good-genes hypothesis. The different mechanisms may be important in different systems.

It's not always the females that do the choosing. In some cases, the females display, and the males choose the females. These examples are very rare, however, so throughout this section, I concentrate on examples in which the female does the choosing. Just remember that in some cases, exactly the same phenomenon occurs, but with the roles of the sexes reversed.

## It's not how you feel, it's how you look: Runaway-selection hypothesis

The runaway-selection hypothesis says that, if females choose mates based on some random showy character simply because they're attracted to that character, that character becomes more pronounced in each successive generation *until* the disadvantages of the having the trait (it makes you less likely to survive) outweigh the advantages (it makes the ladies like you).

Here's how runaway-selection works: Females choose males with, for example, bigger tails for no other reason than that they like big tails. Because the biggest-tailed birds father the most offspring, the average male in the next generation will have a slightly bigger tail. When the next generation's females choose among these males, they select for even longer tails. In this way, male tail length increases until it reaches the point where the increased risk of predation, starvation, or some other unfortunate outcome of having an enormous tail outweighs any additional attractiveness to the females.

Two questions come to mind in considering this model: How does it benefit the females to mate with showy males, and what makes them prefer one particular trait over others? The next sections provide the answers.

The runaway-selection hypothesis seems as though it shouldn't be right simply because . . . well, evolution shouldn't work in such a strange way. But it's important to remember that our emotional reaction to any particular hypothesis rarely has anything to do with whether it's correct.

### Like father, like son: The sexy-sons hypothesis

Beyond looking good as a couple, how does picking a male with a showy tail increase the female's fitness, particularly if the showy tail makes survival more difficult? You'd think that if she wants to get her genes into the future, the female would be better off choosing males with less-showy tails so that her offspring have a better chance of survival.

As it turns out, that's not quite how it works. Instead, specific cases have been found in which female fitness is *reduced* when females mate with less-showy males because the mating success of their sons is lower. In other words, having a sexy son can provide a fitness advantage for a female who chooses to mate with an elaborate male.

Imagine a mutation that makes a female less likely to choose a male with a big tail. Now consider how that choice may affect her fitness. True, her sons wouldn't be hobbled with enormous tails, *but* they wouldn't be prime mating material either. If these poor sons can't get mates, they can't pass on their (and their mom's) genes.

To see whether this hypothesis were true, evolutionary biologists came up with a testable prediction: The sons of less-flashy fathers should have lower mating success than the sons of more-flashy fathers. After identifying species in the wild for which they could measure the attractiveness of males, who mates with whom, and the mating success of the sons of more- and less-flashy fathers, scientists discovered that less-flashy sons actually do have less success in the mating department — and that decreases both their fitness and their mothers'.

### The case for pre-existing preferences

The brain, as you may have noticed, can be a strange thing. It allows us to do a lot of things that are obviously adaptive: find food, avoid bears, and so on. At the same time, it is responsible for traits that don't have obvious advantages, such as dreaming or a propensity to enjoy skydiving. As Chapter 5 explains, not all traits are adaptive. Some traits that aren't adaptive can get dragged along by the traits that are. Pre-existing preferences could fall into this category: in this scenario. females have a pre-existing preference for the trait. Something about how the female brain is wired leads them to favor the trait just because.

Scientists have tested ideas about pre-existing preferences in the laboratory, using fish closely related to swordtails. Swordtails are small freshwater fish common in home aquariums. Males have a long pointed tail — hence, the name *swordtail*. Females, as you've probably guessed, have a mating preference for males with long tails.

The fish used in the studies, although otherwise very similar to the swordtails, differed from them in a couple of key ways: The males lack the long pointed tails, and the females obviously don't choose mates based on tail length, because none of the males has a long tail. To find out what would happen if the males *did* have long tails, researchers attached fake long pointed tails to the male fish. And what did the females do? They preferred to mate with the males wearing the fake tails.

This experiment was important because scientists were able to demonstrate that pre-existing preferences exist. These preferences — in combination with the data showing that the reproductive success of sons is decreased if their fathers were less flashy — suggests that the runaway-selection model is a very real possibility even if the showy trait never had any ecological advantage.

## Or maybe it IS how you feel: The good-genes hypothesis

The good-genes hypothesis proposes that females select mates with elaborate traits because the presence of these traits provides reliable indicators of overall male quality. Under this model, peahens prefer the peacocks with the most elaborate tails because the presence of a large tail indicates that the male has other qualities that would be beneficial for the female's offspring.

For evidence that the good-genes hypothesis sometimes explains female choice, consider a study using a particularly species of stickleback, a small freshwater fish. Males of this species have bright red bellies, and the females preferentially choose to mate with the males with the reddest bellies.

Iain Barber and coworkers measured the resistance to parasitic worms of the offspring of males who differed in how red their bellies were. What they found was that the males with the reddest bellies fathered offspring that were more resistant to parasites. In this case, the redness of a male's belly was a true indicator of his genetic quality with respect to offspring parasite resistance. Females choosing to mate with the reddest males produced fitter offspring than females that selected duller males. The good-genes hypothesis provides a good explanation for female choice in this fish.

## The handicap hypothesis

The handicap hypothesis is a special variation of the good-genes hypothesis. Under the handicap hypothesis, the elaborate male trait indicates the presence of other good genes, specifically because the male trait is so costly that only males with especially good genes could produce the elaborate trait and still survive. According to this hypothesis, a peacock that can survive with an enormous tail must have some other really good genes to balance out the obviously bad tail. Maybe it's especially good at evading predators and at finding food, for example, if it can manage to do both while still dragging around that beautiful but cumbersome tail.

An important component of the handicap hypothesis is that the elaborate male trait must actually be a handicap. In the example of the peacock, researchers have been taking for granted that the peacock's enormous tail hinders its ability to do all the sorts of things that birds need to do to survive.

No studies have been done with actual peacocks (mainly because they don't make good laboratory critters — they're big, they're expensive, and they bite) — but studies of other birds have provided evidence that elaborate tails are a handicap. For their study, Andres Moller and Florentino de Lope used barn swallows. Male barn swallows have long tails, and the females preferentially mate with the males that have the longest tails.

Moller and de Lope artificially increased or decreased the length of male barn swallows' tails and then measured the effects of different tail lengths on feeding ability and mating success. They found that giving males longer tails increased their attractiveness to females and at the same time decreased their efficiency in catching food. These experimental results show that in this bird species, the elaborate trait that is attractive to females is indeed a handicap for the male birds.

# Sexual Selection and Male-Male Competition

As explained in the preceding section, female choice is one type of sexual section. The other type is male-male competition, when males compete with each other for access to females. Competition among males for access to females can take several forms. In each case, natural selection favors characters that increase success in those contests.

In the case of male-male competition, as in female choice, selection favors characters that increase male mating success, which can lead to physical differences between the two sexes. Unlike the case of female choice, however, in which the differences result in males having some increasingly elaborate attractive trait, selection resulting from male-male competition favor traits that facilitate winning contests with other males.

Finally, there may be cases when female choice and male-male competition both operate. For example, male deer use their antlers in contests with other males. Yet it is also possible that large antlers could be appealing to female deer. Female preference could drive the evolution of antler size past the optimum needed to bash other male deer. Or visa versa. Remember that in either case, selection for being able to do all the other more mundane things that deer do (like find food and run away from predators) will place some upper limit on antler size.

The largest deer antlers on record are from the extinct Giant Deer which had antlers 12 feet across! Sexual selection has been suggested as a reason for the evolution of these impressive antlers, but we can't know for sure. If you've heard about this beast, you might have also heard the idea that the reason they went extinct was that their antlers got too big for their own good. But of course, evolution doesn't work that way. Any deer with antlers that were simply too big to allow him to survive wouldn't be passing those super duper antler genes to the next generation.

## *Direct male-male contests*

The most obvious example of male-male competition for females is the case in which males actually fight one another, and the winners are the ones that get to mate. The stakes are high, because the losers won't be passing their genes on to the next generation (at least during this particular mating season), and sometimes the contests are quite violent. Common examples of animals that engage in male-male contests are elephant seals and lions. In both cases, selection has led to increased size in males because larger males are more successful fighters.

### Run away! Run away!

Individuals typically don't engage in contests that they don't stand some chance of winning. For that reason, many contests in the animal kingdom are preceded by a period of posturing in which the two combatants check each other out to judge relative size and strength. After this initial period, a weaker individual chooses to withdraw rather than risk death in a contest he probably can't win. The same logic applies when, part of the way through a fight, one individual realizes that he's beaten and withdraws. This situation is also quite common in nature.

This phenomenon is responsible for male elephant seals being the largest members of the carnivore family. Although female elephant seals are generally between 1,000 and 2,000 pounds, males routinely weigh between 3,000 and 5,000 pounds, with the record weight being more than 7,000 pounds.

Studies have shown that the larger males do indeed father most of the young in elephant seals and some other species. But living the life of a giant warrior-lover isn't all peaches and cream.

Natural selection favors genes that increase size and fighting ability, but not genes for increased longevity. Although enormous size is necessary to fight for a mate, the violent contests and huge energy costs associated with it seem to take their toll. Male elephant-seals live much shorter lives than females — about 14 years as opposed to about 20. Because of how mating is structured in this species, being an older male has no advantage; being a bigger, stronger male does. Death is definitely a bad idea and certainly something to be avoided. But if the only way to get access to females is at the cost of a reduced lifespan, natural selection will favor the bigger-and-stronger genes over the lifespan genes.

From an evolutionary point of view, not reproducing — that is, not passing your genes on to the next generation — is the same as dying. Either way, your genes aren't in the next generation, and your fitness is effectively zero.

## Indirect competition

Competition between males isn't always direct face-to-face conflict. Males also compete in indirect ways. In these cases, selection doesn't favor increased male fighting ability; it favors whatever traits are required to gain access to females.

One obvious example of indirect competition is competition to find females. In many species of butterflies, females mate with the first male that finds them, so finding a female fast is a big advantage. Females often produce chemicals called pheromones that the males can smell. As a result, males have evolved to be incredibly sensitive to these pheromones. Some butterfly males can detect a female more than five miles away!

This type of male-male competition is called *scramble competition* because rather than fighting, the males are scrambling to find the females as fast as they can.

## Sperm competition

In some species, females mate with multiple males, which provides an additional battleground for male-male competition: the battle among the sperm. Natural selection has produced a variety of male adaptations designed to fight this battle, including:

✔ **More sperm:** Males of non-monogamous species produce more sperm than males of monogamous species.

✔ **Seminal plugs:** Most common in insects, but also seen in some vertebrates, is the production of seminal fluid that solidifies into a plug to prevent subsequent males from mating — a strange but predictable adaptation to the problem of competition from other males' sperm.

✔ **Toxic seminal fluid:** Fruit flies are one species that employs this strategy. Their sperm contains toxic compounds that inhibit the sperm of subsequent males. Unfortunately, poison semen isn't too good for the female; head to the later section "The Battle of the Sexes: Male-Female Conflict" to find out more about what happens when what the male wants is at odds with what the female wants.

## Being sneaky: Alternative male strategies

In the evolutionary process, there is often more than one way to accomplish the same task. As biologists continue to study systems in which mating success seems to be determined by the outcome of male-male competition, they're discovering some interesting alternative male strategies. One example is the *sneaky strategy*. Scientists speculate that the sneaky strategy is a way for younger males to have some chance of reproductive success before they get old enough to bark with the big dogs.

Males that use the sneaky strategy avoid direct conflict and try to mate with females while the other males are busy bashing their horns together or otherwise carrying on. It's not clear how common this phenomenon is, but it's been seen in several species, including frogs and red deer. It's also not clear how successful this strategy is compared with the strategy of the dominant male. For males that would be unable to win physical contests, however, it provides the only chance to contribute genes to the next generation.

## The Battle of the Sexes: Male-Female Conflict

In a monogamous species in which males and females enter into long-term reproductive relationships, parents will have an interest in each other's survival because only through the survival of their partners will they be able to produce any offspring. The situation is much different for species with short-lived or no pair bonds. In this case, selection can favor traits of one individual in the pair even if they decrease the partner's fitness.

## A sneaky sperm competition

Several adaptations seem to facilitate the sneaky strategy. One good example occurs in Atlantic salmon. In this species, larger, dominant males guard females and then fertilize their eggs externally as soon as the eggs are laid. The dominant male chases away smaller males that attempt to approach the female he's guarding. The smaller males can't get as close to the female as the dominant male can, but they have an interesting characteristic that may serve to increase their chances of fertilizing at least some of the female's eggs.

Matthew Gage and coworkers found that smaller males' sperm swim faster and can survive longer in the water than those of larger males. Smaller males also produce sperm with different characteristics from those of the dominant males, and these sperm may have a better chance of reaching the female's eggs.

Parents have an interest in providing for the survival of their offspring. From a strictly evolutionary point of view, they don't necessarily have any interest in each other's survival. Why? Because from a fitness perspective, it's better for a male if his partner makes only a few offspring, all of which are his, than if she makes very many offspring that aren't his. The female's fitness is increased by adaptations that favor her reproductive output over that of her mate's. The result of the conflicting goals is an evolutionary version of the battle of the sexes.

The following sections discuss some of the ways this conflict manifests itself in the animal kingdom.

## Infanticide

When male lions take over a pride of female lions, they kill or attempt to kill the cubs. They don't indiscriminately kill baby lions, however. They leave their own offspring alone but kill cubs fathered by the previous dominant male. Killing the cubs sired by the previous dominant male increases the new male's fitness, because it gives him a better shot of producing his own offspring before he gets dethroned by some other male lion.

The violent male-male competition to control a pride of females results in frequent changes in the dominant male. When a new male takes over, he must reproduce as rapidly as possible, as the risk exists that he too will soon be displaced. Because female lions with cubs are not reproductively receptive while they're caring for existing young, the fastest way for the male to gain reproductive access to females is to kill their offspring, which he does as soon as he takes over the pride. The females are then soon ready to mate with him and have his offspring, passing along his genes.

# Should I stay or should I go?

In monogamous species, males and females have either a short or a long-term interest in the fitness of their partners. Non-monogamous species don't have the same concerns. But how would they evolve if they did? Brett Holland and William Rice designed an experiment with fruit flies to investigate this question. They separated the flies into two groups:

✔ For one group of flies, they continued to grow the flies together in large groups in which multiple non-monogamous matings would occur. In this case, an individual fly's fitness is maximized by producing as many offspring as it can, regardless of the consequences for its partners.

✔ In the other group, they raised the flies in monogamous pairs. In this treatment, the fitness of the two flies is interconnected. Flies whose behaviors maximize their partner's fitness also maximize their own fitness.

The researchers raised flies in these two treatments for 47 generations; at the end of the experiment, they measured female fitness characteristics such as life span and reproductive output. Here's what they found:

✔ The male flies in the monogamous pairs had evolved to be less harmful to the female flies. The original male flies were rough; the evolved monogamous ones more gentlemanly.

✔ The female flies had evolved to be less resistant to the male flies. The original female flies watched their backs; the monogamous evolved ones relax and let their hair down.

Their experiment shows that changing the mating system such that the interests of the sexes were no longer in conflict led to rapid evolution that decreased the strength of traits associated with male-female conflict.

In this scenario, the male's fitness increases at the expense of the female's fitness. She's invested time and energy in her litter, which is now lost. Because the new dominant male is so much larger and stronger than the female, she isn't able to protect her young after the new dominant male takes over.

Before then, the females aren't completely powerless, however. Groups of female lions often attempt to chase away solitary male lions that approach the pride. How much — or whether — these actions decrease the probability of the dominant male's being replaced isn't known, but the actions are certainly consistent with the view that the females are trying to prevent the deaths of their cubs.

## Poison semen

Because of the promiscuous mating system of fruit flies, each fly will mate with many other flies over its lifetime. Neither partner has any interest in increasing the fitness of any particular mate, only in increasing its own fitness:

✔ Male fruit flies have toxic chemicals in their seminal fluid that inhibit the sperm of other males (this is an example of sperm competition, explained in the earlier section "Sperm competition" in this chapter). This trait is advantageous for the male because it increases the chance that his sperm will be the ones that fertilize the female's eggs. But the same trait is bad for the female because the semen is toxic to her; she's being poisoned.

✔ The very act of mating seems to decrease the survival of female flies, because repeated mating with males is physically damaging — a situation the female fly tries to avoid.

# Sex: It's Expensive, So Why Bother?

Lots of organisms reproduce sexually, but plenty don't. Many organisms reproduce *asexually,* without sex. Given that sex is expensive and some organisms reproduce just fine without it, why do it? Evolutionary biologists ponder this question because at first glance, sex doesn't seem like a very fit thing to do:

✔ **A sexual organism is effectively throwing away half its genes when it reproduces.** When an organism reproduces sexually, only half its genes get passed to its offspring; the other half come from the other parent. Asexual organisms, on the other hand, pass on *all* their genes. So from a fitness perspective, it would seem that asexual reproduction is better (read, more fit), hands down.

✔ **All other things being equal, the sexual individuals will soon be over-run, and sexual reproduction will be eliminated.** For sex, you need males. Asexual females don't have to make any males, so they produce all daughters, which in turn produce all granddaughters, which produce great-granddaughters, and so on. A sexual female that produces the same number of offspring will make only half as many daughters as the asexual female. She spends the other half of her energy making sons. Because her daughters will, on average, make only half as many daughters as the daughters of the asexual female, the original sexual female will end up with only one quarter as many grandchildren as the asexual female, one eighth as many great-grandchildren, and so on.

But sexual reproduction *hasn't* been eliminated, of course, which means that reproducing sexually must have one or more fitness advantages. So what are these advantages? Well, the truth is that evolutionary biologists just aren't sure. Many ideas have been suggested, and probably more that we haven't thought of yet will be suggested. The following sections look at the current ideas.

The ideas outlined in the following sections are not intended to be mutually exclusive; they could be working together simultaneously to maintain sexual reproduction. The question "Why have sex?" may well have many answers.

---

## Why males don't count

Males don't contribute to population growth; they are needed so that the females can reproduce, but they are otherwise a waste of resources, and that's why they don't get counted. Suppose that two sexual females each make one daughter and one son. The daughter of one mates with the son of the other, and vice versa. Each sexual female ends up with four grandchildren, but that doesn't mean there are eight grandchildren; there are still only four total, and only two of these are female. The population has gone from two females to two females. That's why the males don't get counted.

---

# Idea 1: Sex produces parasite-resistant offspring

Sex produces parasite-resistant offspring, making sexual organisms better able to adapt to changing parasitic environment than asexual organisms are.

Sexual reproduction produces *variable offspring* (offspring that are not genetically identical to their parents), whereas asexual reproduction doesn't. Asexual females produce daughters, granddaughters, and so on that are genetically identical to Mom.

Asexual reproduction is also called *clonal reproduction,* and all the resulting descendants of a single asexual female are referred to collectively as a clone. All the members of a clone are identical except for possible rare random mutations that may have occurred in DNA during the reproductive process. Sexual females, although they produce only half as many daughters, produce daughters that aren't identical to Mom or to one another.

Evolutionary biologists since Darwin have pondered the possible advantages of producing variable offspring. The first argument — that because natural selection requires variation on which to act, having variable offspring makes it more likely that your descendants will evolve faster than the other critters — seems like a no-brainer until you consider the following:

✔ The parent sexual organism clearly has a very fit combination of genes already. It's survived, found food, dodged predators, and found a mate; now it's reproducing.

✔ Sexual reproduction is dicey — you're more likely to break up a perfectly good set of genes with a proven track record in the current environment than hit on an even more fit combination.

So why take the gamble?

The key words here are *current environment*. In the current environment, this sexual organism does just fine. But what happens when the environment changes? This year's very fit gene combinations won't necessarily be very fit next year. The evolutionary interaction between parasites and their hosts can lead to such year-to-year changes in the hosts' environment.

The parasites experience selection pressure to infect their hosts better, whereas the hosts experience selection pressures to better resist infection by their parasites. In this sort of system, the genes of last year's really fit host won't be really fit next year, because the parasites will have evolved to infect it better.

With parasites evolving to better infect their hosts, an asexual host will be in trouble, because all its offspring will be exact copies of itself, but they will have to fend off better-adapted parasites. A sexual host, on the other hand, produces variable offspring with gene combinations the parasite population has not yet had a chance to adapt to. For this reason, the offspring of sexually reproducing organisms are less susceptible to attack by the parasites. So although the parent organism may not fare too well with new or better-adapted parasites, its offspring will.

## Idea 2: Sex speeds up adaptation by combining rare beneficial mutations

Most mutations are bad. After all, randomly changing some piece of an organism's DNA is far more likely to mess up something that was working just fine than it is to improve upon something that wasn't. Every once in a while, however, some random change is actually beneficial and makes an organism better at doing whatever it is that particular organism does. One possible benefit of sexual reproduction is the ability to combine these rare but beneficial mutations more rapidly.

Imagine a couple of beneficial mutations; call them mutation A and mutation B. Both are rare, but they do occasionally occur. Either of the mutations makes the organisms more fit than the organisms without the mutations, but having both is better still.

Sex makes it easier for the two mutations to end up in the same organism. If an individual with mutation B mates with an individual with mutation A, some of the offspring should end up with both beneficial mutations. If this population were an asexual one, each individual would be reproducing in a clonal fashion, and the only way a lineage with mutation A would end up with mutation B (or vice-versa) would be if that mutation occurred in that lineage.

CASE STUDY

# Studying snails and worms

Curt Lively and coworkers tested the theory that sex produces parasite-resistant offspring by using a freshwater snail that lives in New Zealand and is parasitized by a trematode worm. Through their investigation, the researchers provide strong support for the theory that sexual reproduction is advantageous in the presence of parasites.

Lively chose this particular snail because the species has both sexual and asexual forms. The species can reproduce sexually, with a female mating with a male to produce sexual female and male snails. Sometimes, however, offspring have an extra set of chromosomes and are *triploid* instead of *diploid,* which means that they have three sets of chromosomes instead of just two sets. The triploid snails appear to be similar to the diploid snails in every respect except one: They don't need to mate to reproduce. That means that all the triploid snails are females that reproduce asexually, creating identical triploid daughters.

In any given lake, both sexual and asexual snails coexist. The researchers used genetic techniques to determine that many different clones often coexisted in the same location. Each of these clones was the result of a separate instance of sexual reproduction between two diploid snails that resulted in a triploid offspring. Because the different clones have different genes, it's reasonable to assume that one or more of these clones would be better at doing all the things that snails do. The better, more fit clones may outcompete all the others in the short run, but they clearly have not eliminated all the other clones nor the sexual snails.

Lively and company were ready to test some of the specific predictions of the theory that the presence of the parasite was responsible for maintaining sexual reproduction in this system.

**Prediction 1: Sexual individuals should be more common in locations that have more parasites**

Measuring the density of parasites and the frequency of sexual reproduction at many lakes throughout New Zealand, Lively and company found that as the density of parasites increased the frequency of sexual reproduction increased as well. They also measured the frequency of sexually reproducing snails in both shallow and deep areas within individual lakes and found the same thing: As the density of parasites increased, the proportion of snails that reproduced sexually increased.

**Prediction 2: If parasites are adapting to infect their hosts better, parasites should be better at infecting the host snails from their own lake rather than the snails from other lakes**

The researchers performed two sets of experiments to test this prediction. They collected parasites and snails from two lakes on opposite sides of the southern New Zealand Alps and brought them back to the laboratory. (The distance between the lakes made it unlikely that the snails or their parasites had ever encountered each other in nature.) Then they measured the degree to which the parasites from each lake were able to infect snails from the two lakes. They found that for each lake, the parasites were better able to infect the coexisting snails than the snails from the other side of the mountains.

Next, they chose three lakes that were much closer to each other — close enough that ducks could easily fly between them, transporting the parasitic worms from one lake to the next. They again collected worms and snails from the lakes and brought them back to the laboratory to measure the ability of the parasites to infect snails from their own lake as well as the other two. In all three cases, they found that the parasites were better able to infect the snails from their own lake, and the effect was strong enough that it was not overwhelmed by the genetic mixing that could be occurring among the three lakes.

**Prediction 3: The genotypes that were most common in the past should be most susceptible to the current parasites**

In testing this prediction, the researchers focused on the clonal, asexual snails. For their test subjects, they selected clones that possessed the clonal genotypes most common in previous years. Using these clones, they performed two sets of experiments.

They tested the susceptibility of clones from the recent past to the parasitic worms presently inhabiting the lakes from which the clones were collected. What they found was that the clones that had been common in the past were more readily attacked than the clones that had been rare.

If you're a worm, you're well served by having characteristics (naturally selected, of course) that make you better able to overcome the defenses of the most common clone genotypes. A parasite better able to attack these clones would leave more descendents (because there are more of these "common" snails than the others) and therefore would increase faster than those of parasites attacking snails with rare types.

# Idea 3: Sex is beneficial because it can eliminate bad mutations

One of the problems faced by asexual organisms is that, after they have a bad mutation, all their descendents will have the same bad mutation. Then, when a second bad mutation occurs in one of these descendents, all its descendents will have two bad mutations, and so on. This suggestion was first made by Hermann Joseph Muller (1890–1967), and the process is referred to as Muller's Ratchet because the increasing number of bad mutations ratchets down the organism's fitness.

This problem would be especially pronounced in small populations, in which random events might more readily result in the fixation of bad mutations (refer to Chapter 6 for info about how random events impact small populations). Sex provides a solution to this problem. Two sexual organisms, each of which had a different bad mutation, could mate and produce offspring that had neither mutation. Problem solved!

Using a virus that attacks bacteria, called phi-6, Lin Chao and coworkers tested the theory that sex can eliminate harmful mutations.

 Unlike most organisms that have genomes made of DNA, the phi-6 virus has a genome made of RNA. RNA replication is much more likely to result in errors than DNA replication, and as a result, organisms with RNA genomes have much higher mutation rates. This situation makes phi-6 an excellent subject for a study involving mutations, because many mutations occur over a reasonably short experiment.

### Step 1: Producing viruses with decreased fitness

First, Chao established that he could actually observe the phenomenon of Muller's ratchet in the laboratory. He produced viruses that had decreased fitness due to the accumulation of harmful mutations. He did this by making sure that his laboratory population sizes would become quite low periodically. When population sizes are low, evolution via genetic drift can lead to fixation of deleterious mutations, which was what Chao observed. (Refer to Chapter 6 for information about genetic drift.)

Then Chao randomly plucked a few viruses out of one flask to populate the next flask, creating the next generation. (Whereas in a large population, the process of natural selection would weed out less-fit genotypes, Chao increased the chances that a harmful mutation would make it to the next generation by randomly choosing a *few* individuals. In small populations, a virus that's a little less fit still would do fine because it wouldn't have the more-fit strains breathing down its neck.) By repeating this process numerous times, Chao ended up with a population of less-fit viruses on which he could test his idea that sexual reproduction can increase fitness.

### Step 2: Testing whether viral sex leads to increased fitness

With his low-fitness viruses, Chao collaborated with Thutrang Tran and Crystal Matthews in designing experiments to test whether viral sex leads to an increase in fitness. The scientists grew different pairs of the debilitated viruses together with their host bacteria, and they controlled how much sex was going on by altering the ratio of bacteria to viruses. When they put in far fewer bacteria than viruses, they ensured that multiple viruses would be infecting the same host bacterium. As they increased the likelihood of producing viral progeny that had genetic material from two parent viruses, they found that they were more likely to find progeny with increased fitness.

Here's why: An asexually reproducing virus will pass on all its harmful mutations to each of its descendants, but if the descendants have genetic material from two parents, they can end up with the best parts of both. Imagine that one virus has a deleterious mutation on segment 1, and the other one has a deleterious mutation on segment 2. If the viruses are reproducing in the same bacterium, out can pop a virus with neither bad segment.

# Evolution of Separate Sexes and the 50-50 Sex Ratio

Why do humans — and a lot of other species — have different sexes? Why are some individuals males and others females? Evolutionary biologists have offered a couple of suggestions to answer these questions.

## Viral sex

Sex in viruses is quite a bit different from what we commonly think of as sex. In fact, viruses typically reproduce asexually. In phi-6, sex works like this:

1. The virus injects its genome, consisting of three segments, into an unlucky bacterium.

2. The virus hijacks the cell's biochemical machinery to make more copies of itself.

3. Out pop progeny viruses, each of which has copies of the three parental genome segments.

Nothing about this process requires a second virus. Viruses don't need mates and are perfectly able to reproduce asexually. Sometimes, though, two viruses simultaneously infect the same host cell, and all the various replicating bits get mixed together. Now when the new viruses pop out, they can have genome segments from both of the original viruses (that is, one segment from one virus, and the other two segments from the other virus).

These new viruses have two "parents" instead of just one. This process is what scientists are referring to when they talk about viral sex.

Given that two sexes *do* exist, the next question for evolutionary theory is how many offspring of each sex an individual should make. And how is it that species with separate sexes end up with a 50-50 sex ratio: 50 boys for every 50 girls? It turns out that 50-50 sex ratio is something that is perfectly and easily explained by natural selection.

## Sometimes, it's good to be discrete

"Why have males and females?" is a question that often strikes people as surprising, because we humans are so used to having two separate sexes. It doesn't always have to be that way, however. In fact, in many species, the male and female reproductive roles are combined in the same individual! These organisms are called *hermaphrodites,* and if you take a moment to look out the window, you can see that they're just about everywhere.

Most trees are hermaphrodites, containing both male and female functions in the same individual. Each individual apple tree, for example, makes both ovules and pollen. Its pollen is carried to other apple trees to fertilize their ovules while it awaits pollen from other trees to fertilize its own ovules. In the end, all the apple trees produce fruit.

The same is true for most of the fruits you encounter in the orchard or the fruit section of your local supermarket, but not for all of them. Persimmon trees, for example, have distinct sexes. Some persimmon trees are male and make just pollen; others are female and make just the ovules that become the persimmons you buy at the store.

Evolutionary biologists are still trying to figure out the exact mechanisms that led to the evolution of separate sexes. A good place to start is to imagine that increased specialization toward one sex or the other resulted in increased efficiency:

✔ A hermaphrodite has to produce both types of sexual organs, whereas a single-sex individual needs to produce only one set of reproductive organs and can devote the saved resources to additional reproduction.

✔ A single-sex individual can specialize in one particular task, such as finding resources to produce eggs, whereas the other sex can specialize in finding mates.

## One girl for every boy

Because one male is able to fertilize many females, the species will reproduce faster if a higher proportion of females is produced. So the obvious question is "Why not make fewer males?" The reason is one of the key principles of evolution.

Evolution acts most strongly at the level of the individual. The individual isn't concerned with the fate of the population's genes; it's concerned with the fate of its own genes.

An individual's genes are passed down through the generations in this way:

✔ First, the individual makes children.

✔ Then these children make grandchildren.

The 50-50 sex ratio makes sense when you think about the grandchild generation in the following way. Each grandchild has exactly one mother and one father. When the sex ratio of the children is exactly 50-50, the two sexes have identical fitness. If the sex ratio of the children changes, the fitness of the two sexes is no longer balanced.

Imagine a mutation that changes the sex ratio of the children in favor of the production of females. Now more of the children are female than male. As a consequence, the smaller number of males will average more grandchildren than will the more numerous females. Due to the shortage of males, these males end up producing offspring with multiple females. By producing more children, the male offspring has greater fitness. Because of the bias in the sex ratio, they're in a better position to get their genes into the next generation. If there is a mutation that favors producing more males, it will increase until the sex ratio is again 50-50 and the two sexes have identical fitness.

## Changing sexes: Sequential hermaphrodites

An interesting example of the trade-offs inherent in being different sexes is the special case of *sequential hermaphrodites:* species that began life as one sex but then changed into the other sex. In some species, these organisms start out as females and change to males; in other species, they start out as males and change to females. This lifestyle appears in several fish species.

In the case of female-to-male, these species have mating systems whereby males compete for access to females. Large males are more successful than small males and, perhaps as a result of this selection, have favored a system in which small fish avoid being males. All the fish are born females. They mature, mate with the larger males, and reproduce. When they become large enough to compete for mates as males, they simply change sex. Voilà! They grow male sex organs (female organs get reabsorbed) and begin battling to fertilize eggs.

Whenever the sex ratio deviates from 50-50, the rare sex has a fitness advantage, and selection responds by favoring individuals that produce more of the rare sex until the numbers balance out again. As a result, the sex ratio never strays very far from 50-50.

# 'Sex' in Bacteria

To spice things up a bit, I decided to end this chapter with a brief section on sex in bacteria. In this case, I'm talking about sex in the sense of genetic reassortment — that is, the combining of genetic material from different individuals.

Bacteria don't need sex to reproduce. For the most part, they reproduce by dividing in half. One bacterium makes two bacteria; then these two make four, and so on. But occasionally, a bacterium acquires DNA from other bacteria and combines that DNA with its genome; henceforth, all its descendents have this new combination of genes.

Here are a few interesting tidbits about the product of bacterial sex:

✔ **Bacteria don't seem to be very picky about which bacteria they get the new genes from.** Some bacteria even absorb free DNA from the environment and incorporate these genes into their own genomes. This phenomenon is especially puzzling from a fitness perspective, because the free DNA has most likely come from bacteria that died and ruptured, releasing their genetic material into the environment. Because these particular bacteria died, it's not clear why other bacteria would want their genes.

> ✔ **Bacteria occasionally incorporate large numbers of genes from distantly related bacteria.** A trick like this wouldn't even be possible for most organisms, because even if they had a way to transfer some of the genes — from a pine tree to a duck, for example — the resulting combination wouldn't actually be functional. Most organisms have tightly interconnected sets of genes, and it's not possible to just splice into a genome a bunch of genes from some other organism and expect the resulting combination to work. You probably wouldn't end up with a functional organism.

Although scientists don't completely understand the selective forces that favor bacteria's ability to incorporate foreign genes, they have the tools to see its results. With the increase in the use of DNA sequencing — the technique that determines the specific sequence of an organism's DNA — scientists have discovered that the new genes the bacteria acquire often confer new function.

Take the common intestinal bacteria *E. coli.* You have some of it living in you right now. It keeps to itself and doesn't do you any harm. It's even possible that by taking up space, this particular strain of *E. coli* keeps more harmful bacteria from taking over your intestines.

But some strains of *E. coli* don't just sit quietly in your gut; they disrupt the intestinal walls and cause illness. *E. coli* O157:H7, commonly known as Jack in the Box *E. coli,* is one such strain, and it's an especially nasty one. The reason this particular *E. coli* strain causes illness is that it has picked up a variety of toxin-producing genes from other bacteria, including a large cluster from the bacterial species Shigella, which causes dysentery in humans and animals.

*E. coli* O157:H7 was first noticed in 1982, and the Centers for Disease Control and Prevention estimates that it causes approximately 73,000 cases of illness and 61 deaths in the United States each year.

Scientists don't have a clear understanding of what regulates bacterial sex. But at least in the case of E. coli O157:H6, the new combination of genes seems pretty good from the bacteria's perspective. O157:H7 has been extremely successful. Today, it can be found in most cattle farms and most petting zoos. So remember to wash your hands regularly!

# Chapter 13

# Co-evolution: The Evolution of Interacting Species

*C*o-evolution is what happens when a change in one species selects for a change in another species, and it's ridiculously cool because it makes for a lot of fascinating species interactions. When two species are evolving reciprocally, they end up with some really beautiful patterns and neat adaptations. I mean, think about it — cheetahs run ridiculously fast. Why? Because the antelope that they chase run really fast too. The existence of cheetahs selects for faster antelope, and those faster antelope select for faster cheetahs.

The cheetah-antelope scenario is an example of an *antagonist interaction,* in which one species is looking to eat the other one. Possibly even more beautiful are complicated examples of *mutualism,* in which, for example, plants with long curved flowers are pollinated by hummingbirds with long curved beaks — beaks that are curved to match the curved flowers exactly!

But even beyond how amazing all these interactions are, co-evolution serves as a nice reminder that evolution is always about your own fitness, not somebody else's.

## Co-evolution Defined

Co-evolution is what happens when interacting species evolve together. A change in species A selects for a change in species B, which then selects for another change in species A, which in turn selects for another change in species B . . . and so on and so forth, ad infinitum.

For co-evolution to occur, the interacting species must affect each other's survival and reproduction — in other words, their fitness. The antelope affects the cheetah when it avoids being eaten; the cheetah affects the antelope when it eats it. By running away, antelope affect which cheetah genes end up in the next population (hint: the fast ones). Cheetahs affect which antelope genes end up in the next population (yep, the genes for fastness).

Having said that, I ask you to keep in mind that the strength of these interactions doesn't have to be equal. A cheetah that just misses an antelope misses lunch; it might get one tomorrow. An antelope that just misses getting away *is* lunch; it has no tomorrows. Hence, the selection on antelope by cheetahs may be a little stronger than the selection on cheetahs by antelope.

The following sections explain the types of interactions that co-evolving species can have and the outcomes that co-evolution can result in. The remaining sections of this chapter offer examples of co-evolution in nature and in the laboratory.

## Co-evolution and species interactions

Co-evolution requires that two or more species interact such that they evolve in response to each other. Sometimes, these interactions are beneficial to both parties, sometimes they're beneficial to just one and bad for the other, and sometimes they're bad for both.

The following sections explain the main categories of species' interactions.

Although these discussions focus on pairs of species, co-evolution can involve interactions among more than two species. The relationship categories remain the same, regardless of how many organisms are co-evolving, even though any one species may be interacting with other species in a variety of different ways (like trying to eat some and not get eaten by others).

### Mutualism: You scratch my back; I'll scratch yours

In a mutualistic interaction, the presence of each species has a positive effect on the other. Bees and flowers, for example, can co-evolve mutualistically. Another example is the sea anemone and the clown fish. The clown fish hangs out (and even lays its eggs) around the poisonous tentacles of the anemone, but doesn't get stung. Why? Because of a combination of the clown fish's sting resistant mucus and the fact that the anemone doesn't mind it being there.

Every once in awhile the clown fish will leave the safety of the stinging tentacles and venture off into the surrounding waters where its bright colors make it visible to predatory fish. When the predatory fish attacks, the clown fish

## Commensalism

For co-evolution to occur, both species have to be affected. For that reason, commensalism — in which the interaction affects only one of the partners, not both — isn't really an example of co-evolution, but it's still interesting.

In *commensal* interactions, one species receives a benefit, while the other remains unaffected. Little organisms that hitch rides on bigger ones or use them as places to live are examples.

No evidence exists that a turtle cares whether a little bit of algae grows on it, but the turtle gives the algae a nice place to live. But if being on a turtle is an important part of algal ecology, there might be selection favoring algae that better stick to turtles.

heads back into the anemone, the predator pursues, and the anemone eats the predator, leaving the scraps for the clown fish! Clown fish that lure prey into the anemone tentacles get more food for themselves while anemones that provide a safe haven for clown fish get more food, too. The participants are in it for themselves, but the result is a beautiful mutualistic interaction.

Here's a key point to keep in mind: The connotation of the term *mutualism* and the fact that both of the co-evolving species benefit may lead you to think that the species *intend* to help each other out. That is absolutely not the case. Each organism does what it does for its own benefit. The fact that it's also benefiting the other organism is irrelevant.

### Competition: Unrest in the forest

In a competitive interaction, each species has a negative effect on the other. Take trees, for example. Why are trees tall, and why do they have trunks? The energy that trees convert from sunlight needs to travel all the way down to the roots, and the water from the roots needs to travel all the way up to the leaves, but the trunk is expensive to make and not the most efficient conveyor of nutrients or energy. It certainly seems that a more-efficient arrangement would be to have the leaves close to the water supply instead of many feet off the ground.

The answer is that trees have to fight for light — a key ingredient in their fitness. A tree can survive only if it's tall enough not to be overshadowed by the tree next to it. When you have to grow tall, a stem just won't cut it. You need something a bit more substantive — hence, the trunk. The next thing you know, all the trees have to make trunks; otherwise, they're overshadowed by the trunky trees next to them.

There's a limit to how tall a tree can grow, of course. This limit is determined by the amount of light that hits the forest where that particular tree grows, the amount of rain, and the soil condition and type. These factors limit how much energy a tree can devote to making trunks. Another limiting factor is wind, which may blow a tree down when the tree is too tall.

### Predation and parasitism: I'm a giver; you're a taker

In both predation and parasitism, one species has a positive effect on the second species, but the second species has a negative effect on the first. Examples are predators and parasites. Antelope are good for cheetahs, but cheetahs are bad for antelope. Ditto a dog and its fleas.

The antelope-cheetah interaction is an example of predation. The dog-flea interaction is an example of parasitism. Whether the system is one or the other depends on whether the parasitic organism wants to eat all of you or some of you. In some parasitic interactions — such as deadly parasitic diseases, to which you succumb in the end — there's not much fitness difference being eaten by a cheetah or wasting away from a disease. You're dead either way. You can read about co-evolving disease organisms in "Diseased Systems: Parasitic Co-evolution," later in this chapter.

### When the interactions change

As you think about co-evolution, keep in mind that these interactions aren't always completely fixed in nature. The type of co-evolution between species can change over time. Bees, for example, need nectar; flowers need pollinators. The two species co-evolve in a mutualism: Their interaction is mutually beneficial because when the bee takes the nectar, the flower gets pollinated.

Now imagine a mutation in bees that results in their boring holes in the side of flowers and sucking up the nectar. The system has gone from being mutually beneficial to being parasitic: One species benefits to the detriment of the

## The pronghorn antelope

Evolution has an endless number of fascinating stories, and here's another one for your reading pleasure. The fastest land mammal in North America is the pronghorn antelope, which can outrun by a substantial margin absolutely everything that it might ever come across. Why on earth does it run so fast? How could natural selection have caused that speed if nothing is chasing this antelope?

Well, as it turns out, even though no cheetahs exist in North America today, plenty of cheetah fossils turn up on the continent. I can't say for sure that those fossil cheetahs used to chase pronghorns, but they did chase something, and they ran very fast (info gleaned from their bone structure being similar to that of modern cheetahs in Africa).

other. The flowers' nectar is gone, and no pollination has taken place because the bees aren't coming anywhere near where the pollen is or needs to be.

In response, natural selection will favor plants that keep the bees from getting the nectar. Selection may make the flower harder to chew through, in which case plants that happen to have thicker flowers are going to leave more descendents because they still have some nectar left in them to attract other pollinators. The process is still co-evolution, but a transition from one type of interaction to another has occurred.

# Outcomes of co-evolution

In co-evolution, one organism evolves to its own benefit in response to the other organism. As one changes, the other changes. So what's the endpoint for all this change?

In antagonistic interactions, the end point could be that one organism eliminates the other. In mutualisms, the end point would be when there's no further selection for a tighter association. In the example of the long-flowered orchid and the long-tongued moth, unless something changes to make the association less or more beneficial for either organism, the flower's length and the moth's tongue are plenty long enough. Other times, the end point is just that an organism has reached the limits of how tall or fast or whatever it can be.

## The Red Queen

One possible outcome is a scenario referred to as the Red Queen hypothesis. The expression comes from *Alice in Wonderland,* in which the queen runs in place but doesn't get anywhere.

In the Red Queen scenario, species that co-evolve essentially run in place. You can find lots of examples in the co-evolution of plants and insects. When a plant evolves a novel chemical defense mechanism, for example, it drives the insects to evolve a novel detoxification mechanism. You can also see the Red Queen scenario in the fossil record. Fossil evidence shows steady advances in characteristics such as shell thickness and brain size in predators and prey.

You can think of co-evolution acting this way if you imagine it in the context of an arms race. To give itself an edge, each side evolves new or better adaptations, which the other side counteracts as it adapts in response. At the start of WWI, for example, pilots shot at each other from the cockpit with pistols. Frightfully fearsome — until somebody strapped a whole machine gun to the front of a plane. Reciprocal adaptations of this sort took aircraft from the wood-and-fabric aeroplanes that the Red Baron flew over the fields of France to the carbon-fiber-and-titanium jets that fighter pilots "strap on" today. Where does that leave the respective air forces? Traveling faster than the speed of sound but not really much ahead of the other guy.

### Extinction

Another possible outcome of co-evolution is that one of the participants can't keep up and goes extinct. Obviously, scientists don't see these systems, because one of the participants no longer exists, but they know from the fossil record that extinction is a common phenomenon. Lots of organisms that used to be around just couldn't keep up.

### A little stability

Yet another possible outcome is that the process leads to some, at least temporarily, stable end point. An example would be cheetahs and antelope whizzing across the plains of Africa at 65 miles an hour. Given the physics of the mammalian body plan, animals are unlikely to evolve to run at the speed of sound. The cheetah is at the current upper limit of animal land mammal speed. (Scientists don't know how much faster an animal could run, but they know that some upper limit must exist.)

# Interactions between Plants and Animals

Some of the best examples of co-evolution involve interactions among plants, animals, and insects. These interactions revolve around:

- **Sex (more specifically, reproduction through pollination or seed dispersal):** Plant reproduction relies on getting pollinated or sowing seeds. The flowers of some plants, for example, require particular pollinators; not just any animal or insect will do. In this case, the co-evolutionary pair — the plant and the specific pollinator — evolve to maximize their own fitness, each trying to get more of what's good for it from the other.

  Other plants don't require a specific pollinator; they share pollinators with other plants. In these cases, the plant species have to compete, so natural selection favors traits that make each species better than the other at attracting pollinators.

- **Protection in exchange for room and board:** In some systems, plants provide food and shelter to insects in return for the protection that the insects provide.

Although these instances certainly aren't the only examples of co-evolution, they offer a good glimpse of the way interactions among species can influence the characteristics that the species evolve. Also, they let me write about a moth that has a 10-inch-long tongue.

# Pollination wars

Pollination involves plants and often insects. (Forget pollination by other animals — such as birds and bats — and wind for now; those situations aren't important in this discussion.) Each species has a vested interest in getting what it needs. The plant, for example, is trying to get its pollen to (or get pollen from) another plant of the same species. To do that, the plant needs to spend effort and resources on attracting pollinators. It has absolutely no interest in helping its pollinator unless, by doing so, it helps itself. For its part, the insect (or pollinator) cares nothing about whether the plant gets pollinated; it's after the reward offered, or promised, by the plant.

Ideally, plants want a pollinator that's very *species-specific* — that is, attracted only to plants of the same species. After all, what good is a pollinator that dumps your pollen on incompatible plants? With a species-specific pollinator, the plant would be assured that its pollen would go to an individual of the same species.

Consider the orchid species that has nectar 10 inches down in a very long, thin flower. Based on his study of this orchid, Charles Darwin predicted that the pollinator would be a moth with a 10-inch tongue. At the time, no one had ever seen a moth with a tongue that long, and people considered his idea to be pretty ridiculous. But 40 years after Darwin's prediction, exactly such a moth was found — and whaddya know, it turned out to be the pollinator of this particular orchid. In honor of Darwin's prediction, the moth was named *Xanthopan morgani subspecies praedicta,* just barely beating out *Darwin He's-the-Man Moth.*

The example of the moth with the long tongue shows the benefits of a tight association between the two mutualists. The orchid gets its pollen delivered only to its own species while the moth gets access to resources that the shorter tongued moths can't reach. But there are also potential downsides to being involved in such a tight interaction:

- If one of the partners is absent, then the other is out of luck. And if one species should go extinct, the other one might not be far behind.

- It reduces the probability of dispersal to new environments. Imagine an orchid seed carried on the wind to a remote island. The soil's just right, the temperature is perfect, and lots of other plants are doing just fine. But that orchid had better hope that some of those moths get blown over to the island too! Because if none do, then that wandering orchid is doomed.

# The evolution of pollination by animals

How animal pollination evolved in the first place is fun to think about. Pollination by animals has been an incredibly successful strategy for plants. Not that wind pollination doesn't work, but it has its disadvantages, first of which is the fact that the plant has to make a huge amount of pollen, because most of it isn't going to get anywhere near the target.

An interesting feature of wind pollination is that the plant needs to be able to snatch the pollen from the air. Toward this end, some plants have developed sticky fluids that assist in pollen capture. Some scientists envision that animal pollination arose when insects or other animals fed on these plant secretions and, as a result, moved pollen from one plant to another.

### Figs and wasps: I couldn't live without you, baby

One of the classic examples of a mutualistic interaction is that between figs and the wasps that pollinate them.

The fig's reproductive structure is complicated. The flowering portion of the fig is a hollow structure with male flowers on the outside and female flowers, which produce seeds, on the inside (which is called the *syconium,* but I'm guessing that you don't care!). Here's what happens:

1. A female wasp arrives at one of these structures and gathers pollen. She then enters and deposits pollen on the female flowers within and also deposits an egg in each of the ovules that she can reach. Then she dies.

2. The eggs hatch; the larvae feed on the developing seeds; the larvae pupate; and adult wasps emerge.

3. The males mate with females and chew a hole through which the females depart. The male promptly dies; lacking wings, he wasn't going anywhere anyway.

So how is it good for the plant to have eggs dumped inside it, its developing seeds eaten, and a hole chewed through it? As it turns out, the female flowers on the inside are different lengths. Some are too long for the wasp to reach down and deposit an egg on. As a result, some seeds escape predation by the developing wasps. So the wasp gets food for her offspring, and the fig gets a very reliable pollinator.

Even though in this case the plant and the insect are completely reliant on each other for survival, each species is acting in its own interests. The fig wasps don't pollinate the fig tree because they care about the fig; they do it because their offspring will feast on the developing seeds. And successful fig trees are the ones that produce seeds above and beyond the ones that get eaten. The wasp uses the fig; the fig uses the wasp; and the species are so tightly co-evolved that each would go extinct without the other (what scientists term an *obligate mutualism*) because each species of wasp pollinates only a single species of fig, and vice versa. But they're still not pals!

Oh, and if you've been keeping track of the dead wasps, you've noted that every time you eat a fig, you're eating a few dead wasps, too. Don't worry, though; they're very tiny — small enough to fit through the eye of a needle.

### Your cheatin' heart: Non-mutualism pollinators

Some plants don't play fair. They cheat to get what they want and leave the poor, gullible insects no better off for their trouble. Basically, the plant tricks the pollinator into visiting the flower with the promise of a good time or a good meal and then doesn't pay up. As you can imagine, the interactions between these plants and their unsuspecting pollinators are not mutualistic; one of the partners is coming out behind. In these cases, co-evolution acts to make the plant better at deceiving the pollinator and the pollinator better at not being deceived.

A couple of examples:

- Some plants have flowers that smell like rotting meat. These flowers are pollinated by flies and beetles that arrive expecting to find a carcass. Instead, they end up getting a dusting of the flower's pollen, which they deposit the next time they make the same mistake.

- Several species of orchids have evolved the ability to attract male insects by mimicking the sex pheromones, and sometimes the shape, of the female insect. Imagining what happens in this case is easy: The male insect pounces on the orchid and makes a really good attempt at mating with the flower, getting covered with pollen in the process. In the end, he gives up because the process just isn't going very well and flies off to find a different female. But he may end up on another orchid and transfer the pollen that he picked up on the first orchid.

  Don't blame the poor male insects for being clueless. The compounds that the female insects use to attract males are actually very similar chemically to the waxy substances that the orchid uses to avoid drying out. You can easily imagine how at some point, one or a few mutations resulted in an orchid that smelled good in a whole different way.

### Parasitic non-pollinators

Just as some plants avoid giving a reward, some animals take the reward but don't provide any pollination services. Think back to the 10-inch-long flower. Many flowers have deep (though not quite so extreme) nectar sources, which make them inaccessible to insects with short proboscises.

Some insects have developed a clever solution to this problem: They just hold the base of the flower, chew a hole, and slurp out the nectar, never getting anywhere near the pollen. In this case, the insect is a parasite of the flower. As the species co-evolve, the plant species is selected to have less-chewable flowers, and the insect is selected to be better able to chew its way to the nectar.

### Sharing pollinators

Plants that share pollinators have to compete to attract them. This competition among plants introduces another place for selection to operate. Plants that rely on the same pollinators, for example, will be selected to flower at different times.

By flowering in a narrow window of time, a plant increases the chance that its pollen will go to an individual of the same species. Similarly, a species of plant that flowers at a time when other species don't has better access to pollinators and, as a result, is more likely to leave descendents.

## Seed dispersal

If seeds are light and small, they can be dispersed by wind. The problem with this system is that the plant has to make a lot of seeds, because most of them end up in the wrong place.

Having an animal disperse your seeds can solve this problem, though it may introduce others (the animal might eat your seeds). Animals can move larger seeds, and they provide a bit more specificity in where the seeds end up. If a particular bird is eating a particular fruit, that bird is known to visit places where that particular kind of plant can grow and may end up depositing the seeds in a similar habitat. This system is how mistletoe seeds get dispersed. Mistletoe is a parasitic plant that grows on other plants. Mistletoe fruit is eaten and the seeds dispersed by perching birds, which conveniently deposit the seeds right onto another potential host plant. The fruit is the lure that gets the birds to eat the seed.

Not all relationships that result in seed dispersal are mutualisms, of course. Some plants are very good at getting animals to disperse their seeds without giving any reward. If you've ever had to pull seeds out of your socks or off your dog after a hike, you know exactly what I'm talking about. Some plants have evolved quite clever mechanisms that allow their seeds to be attached to animals.

Not all seed or fruit eaters disperse the seeds; some just digest those as well, setting up another co-evolutionary interaction that has resulted in plants producing seeds with toxins and animals developing the ability to cope with these toxins.

## Trading food and shelter for defense

This mutualism is my favorite one, occurring between the bullthorn acacia (*Acacia cornigera*) and the ant *Pseudomyrmex ferrugiea.*

The bullthorn acacia has large hollow thorns where ants live, and it produces nectar (not associated with flowers) and protein-rich structures that the ants eat. The ants patrol the plant and sting anything that tries to nibble or land on it, and they clear the ground under and around the plant, killing any plants that might compete with the host acacia.

Not such a big deal, right? Wrong. The bullthorn acacia is unusual. Other acacias don't produce extrafloral nectaries; they don't produce the protein structures, which have no known function other than to feed the ants; and they're poisonous, producing chemicals to defend against herbivores. The bullthorn acacia doesn't have these characteristics because it doesn't need them; it has the ants instead.

And let me tell you, these ants really hurt when they sting. I know because I checked — personally. And I'm not the only person who thinks so. The Schmidt Sting Pain Index — yes, there really is one; you can see for yourself by going to `http://scientiaestpotentia.blogspot.com/2006/06/schmidt-sting-pain-index.html` — rates them at 1.8 (a bald-faced hornet gets a 2) and describes the pain as "A rare, piercing, elevated sort of pain. Someone has fired a staple into your cheek." Someone there must have been stung, too, which just goes to prove the lengths to which scientists will go to collect data.

# Disease Systems: Parasitic Co-evolution

In host-parasite systems — specifically, disease systems — one member of the co-evolving pair preys on the other. In such a system, scientists expect the host to evolve increased resistance to the parasite. All things being equal, being resistant must always be better.

So how should scientists expect the disease organism to evolve? A common misconception is that selection will cause the disease organism to evolve in a way that's less harmful to its host. The rationale: Because the disease organism needs the host, it should be nice (i.e., less virulent) to it. (*Virulence* refers to the fitness decrease that results from infection with a particular disease. A disease that kills you right away has a high virulence; if it just gives you the sniffles, and you can still make it to the office, it has a low virulence.)

But that's not the way it works. To a disease organism, maximum fitness isn't just about surviving; it's also about spreading to other hosts. Therefore, evolution selects for whatever virulence level makes the disease more effectively transmitted. A disease will become highly virulent, for example, when high virulence increases its fitness even as it decreases its host's fitness.

The specific virulence that maximizes fitness varies from organism to organism, based on how the disease is transmitted. This point is key. No one "right" level of virulence exists. The common cold won't be more fit if it kills me before I can make it to school, but when I'm at school, it's more fit if it makes me cough on someone else.

# Bunnies in the Outback

Australia didn't have any native rabbits. A few were introduced in the mid-1800s, and by the mid-1900s, Australia had half a billion. That's a lot of rabbits. In 1950, the myxoma virus was introduced to control the rabbit population. The virus, which had an extremely high fatality rate, was very effective. But soon after it was introduced, the virus evolved to be less virulent.

Here's why: The virus was spread from rabbit to rabbit via mosquitoes, which bit infected rabbits and transferred the virus to uninfected rabbits. It just so happened that if the host rabbits had survived longer, a greater chance existed that a mosquito would bite an infected rabbit and transfer the virus to a new host.

That being the case, you'd think that the virus would eventually evolve extremely low virulence to maximize the opportunity for transmission. But this didn't happen either, because mosquito transmission is maximized if the rabbit is covered with virus-filled lesions on which mosquitoes can feed. (If that situation sounds like it's bad for the rabbit, that's because it is.)

The myxoma virus needs a certain level of virulence to be transmitted to a new host. It's not good for the virus to kill its host immediately or to float around in the bloodstream relatively harmlessly. But it *is* good for the virus if the host is covered with lesions.

The rabbit population wasn't taking this virus lying down; it was evolving to be more resistant to the virus. As the rabbit population became more resistant, the virus population evolved increased virulence to maintain maximum mosquito transmissibility.

Scientists know all these things because the original rabbits (from places other than Australia that had never been exposed to the myxoma virus) and the original virus were still available for study. Using the original virus, scientists showed that the rabbits living in Australia evolved increased resistance. Using the original rabbit population, scientists showed that the virulence of the virus in the Australian rabbit population first decreased but then increased in concert with the increased resistance of the host rabbit population. So there you have it — a classic example of host-disease co-evolution in a natural setting.

## Disease-host interaction in the lab

In trying to understand disease-host interaction, Sharon Messenger, Ian Molineux, and J.J. Bull conducted an experiment to examine how different viral transfer mechanisms affect the relative advantage of different levels of viral virulence. The two basic types of transfer mechanisms are

- **Vertical:** Transmission from parent to descendent (essentially, transmission through time)
- **Horizontal:** Transmission from one individual to the next in the current time

What they discovered is that when many hosts are available, a virus that harms its host but increases the chance that its progeny will infect new hosts is favored. But when few hosts are available (that is, a reduced chance of horizontal transfer exists), selection favors a virus that's less harmful to its host, because only through the host's survival and reproduction can the virus survive and reproduce.

The experiment used bacteria and a virus that infects the bacteria. The experiment first selected for different varieties of the virus that had a range of effects on host growth rate. More-benevolent strains had a small effect on host growth rate but couldn't be transmitted horizontally, and less-benevolent strains had a larger effect on host growth rate and could be transmitted horizontally. This part of the experiment generated a variable virus population. After the experimenters had variation of exactly the sort that interested them in their virus population, they were able to set up different experimental scenarios to see when the different viral variants would have higher fitness.

Taking a 50-50 mixture of bacteria infected with the two viral types, the scientists grew them in the presence or absence of additional bacteria that either were or weren't susceptible to infection. After allowing time for viral and bacterial reproduction, they assessed the relative proportion of the two viral types at the end of the experiment. As predicted, in the absence of an opportunity for horizontal transfer, lower virulence was selectively favored. In the presence of available susceptible host bacteria, the less-benevolent viral strain was favored even though it was more harmful to its host.

# Chapter 14

# Evo-Devo: The Evolution of Development

. . . . . . . . . . . . . . . . . . . . . . . . . . . . . . . . . . . . . . . . . . . .

## In This Chapter

▶ Developing an understanding of development

▶ Discovering how a little change can make a big difference

▶ Finding the deep similarities between very different animals

. . . . . . . . . . . . . . . . . . . . . . . . . . . . . . . . . . . . . . . . . . . .

**A** species with one particular form can give rise to a species with another form. Take humans and fruit flies, for example. They have a common ancestor somewhere in the distant past. This common ancestor gave rise to two very different organisms. The question is how. What process is at work that results in such different creatures?

The answer has to do with the interplay between evolution and development, which is one of the hottest areas of current evolutionary research and one in which today's scientists are able to learn lots of things that people back in Charles Darwin's time didn't have a clue about.

This chapter talks primarily about the development of animals because animal development is for the most part *deterministic* — that is, all the members of the species end up looking pretty much identical, at least structurally. Humans have two arms, two legs, one heart, one head, and so on, and all these parts need to be in the right places. Compare this structure with that of a maple tree; branches can go every which way, and their reproductive structures (flowers) can be all over the place. Lop off one of the branches, and another one may grow and produce its own flowers — definitely not the way that human structures work! All this doesn't mean that plants don't have development; they're just a little bit more free-form about it.

# Defining Development: From Embryo to Adult

*Development,* in evolutionary terms, refers to the process by which a fertilized egg develops into the adult form.

By *adult form,* I don't mean that the organism has reached the age of majority and is heading off to college — or, for that matter, has reached sexual maturity and is getting ready to set up its own pride on the Serengeti. In evolutionary terms, development ends when the organism has all its parts, that is, when it has its final shape. In this context, *adult* doesn't mean *grown up*.

The time frame for development differs for each organism, but the process starts with a single cell, which develops as a result of

- ✔ **Cell division:** The process whereby one cell splits into two, two into four, four into eight, and so on.

- ✔ **Differentiation:** The process whereby one lineage of cells gives rise to different types of cells of another lineage, such as a skin cell or a heart-muscle cell, or any number of other types of cells.

The *embryo stage* starts with the first cell division and goes until the organism has all its adult parts. This stage doesn't end at birth (most animals aren't born, but develop from eggs outside the mother) but at some time before birth, when all the parts that will be recognizable in the adult organism have formed. The organism has legs, eyes, all the various internal organ systems, and so on. As you can imagine in complex organisms such as mammals, pinning down exactly when the embryo stage ends is difficult. The key is that you know it when you see it: The developing embryo looks like a little version of the adult organism.

Humans have a lot more names for various stages of development. In human reproduction, you hear the terms *zygote* (to refer to the fertilized egg), *embryo* (to mean any stage of development before birth or the particular early stage of development), *fetus* (to refer to later developmental stages), and *pain in the keister* (to refer to the teenage years). Don't let these alternative uses confuse you. In this chapter, the term *embryo* refers to the developmental stage from the first cell division to the formation of all the adult parts.

## Under construction: The development process in action

Starting from a single cell, the embryo divides and grows. As this growth progresses, different lineages of cells become specialized to perform different tasks. All the cells in the organism contain the same DNA — the same instructions, but the instructions are expressed differently in the different cells. In this way, the various structures of the organism develop.

From initial populations of cells that haven't yet specialized (called *pluripotent cells*) and that can transition into any cell type, specific cell lineages are derived. These pluripotent cells are called *embryonic stem cells.* Some will give rise to skin cells, others to bone cells, and so on.

After cells have transitioned to specific cell types, the lineage's future appears to be fixed. Skin cells divide to produce other skin cells, for example; they can't make other types of cells. Liver cells grow, divide, and go on to form the liver; they don't go popping up in other parts of the body.

Scientists have made great strides in understanding how this process works, and I go into some of the details in the next sections. But for now, keep two things in mind:

- Starting from cells with exactly the same DNA, it's possible to obtain cells of very many types. One component of the development process is the mechanism by which this differentiation occurs.

- To make an organism, the different cell types have to develop in the correct place. The spatial patterning within the developing embryo is a key to creating a viable organism. (The liver needs to develop in the abdomen, for example, not in the skull!)

## The effect of environment

An organism's *phenotype* (physical characteristics) is a result of the interaction between its *genotype* (genetic makeup) and the environment. A person with the genetic potential to be 7 feet tall won't achieve that height in the absence of proper diet, for example. A malnourished person will be stunted compared with a genetically identical individual who was well fed. The impact of the environment on the developing organism is called *environmental effects,* and one way that the environment can affect phenotype is by affecting development. (You can read more about what affects phenotype in Chapters 4 and 7.)

The following sections highlight examples of environmental effects. Here's the take-home message: Small changes in the regulation of development in genetically similar individuals can have major effects.

### The development of queen vs. worker bees

A beehive consists of at least one queen bee who lays the eggs, a larger number of worker bees who tend the eggs, and developing larvae. The queen and worker bees look very different. The most noticeable difference is the size of the queen: she's much larger than the workers. Yet the size difference isn't the result of different genes.

Whether an egg develops into a queen or into a worker bee depends on its environment within the hive, specifically whether it's fed exclusively royal jelly for the first days of its life. Royal jelly is a substance produced by special glands of worker bees and fed to all larvae. An all-royal-jelly diet equals a new queen.

The existence of different castes in bees (and in other social *hymenoptera* — a class of insects that includes ants and wasps) offers a nice example of how changes in the path of development can result in organisms with different body forms. We know that any female bee can become a worker or a queen — the final form isn't based on differences at the DNA sequence level but instead on differences in how those genes are expressed.

Researchers can look at patterns of gene expression in developing bees and figure out exactly which genes are regulated differently in queens versus workers. Queens develop from eggs that experience an increase in the production of metabolic enzymes and that regulate the genes associated with hormonal activity differently.

### The Thrifty Phenotype hypothesis: Genes for flexibility

For mammals, the embryo develops within the mother; therefore, the maternal environment influences development. Because one major component of the maternal environment is how well nourished Mom is, embryonic development may respond to changes in maternal nutrition — a position that medical evidence seems to support.

Medical evidence indicates that a fetus deprived of nutrition during key parts of development will develop into a baby with a greater degree of *metabolic thriftiness,* a group of characteristics that reduce caloric requirements (smaller size and lower metabolism, for example).

Some scientists postulate that these fetal changes may be the result of natural selection; genes that allow a developing human embryo to better prepare for existence outside the mother will be selectively favored. This hypothesis is called the *Thrifty Phenotype hypothesis,* and it goes something like this: Some of the fetal developmental changes observed in low-birth-weight babies may have been adaptive early in human history. If the mother's condition predicted low food availability when the baby was born (for example, she was deprived of food during pregnancy), genes that allowed for developmental flexibility — like growing slowly, but more efficiently, when food is scarce — may be selectively advantageous.

From an evolutionary viewpoint, what the fetus would be responding to is unclear. The condition of the mother could correlate with any of the following:

✔ The environment in which the mother was living

✔ The mother's ability to provide resources

✔ The environment in which the adult offspring would find itself as predicted by the mother's environment (in those situations in which the environment changes slowly over a time scale longer than the organism's generation time)

But the jury's still out on whether natural selection had anything to do with the fact that human embryos may develop differently when maternal resources are scarce. At this point, the Thrifty Phenotype hypothesis is pure speculation (often the first part of scientific inquiry). What's not speculation is that embryonic developmental changes do occur, which unfortunately are related to other medical problems, such as diabetes and obesity, so being able to identify the cause and the mechanism by which the development pathway is altered is medically important.

The thrifty phenotype hypothesis postulates genes for *plasticity,* genes that allow the developing fetus to develop to be more or less thrifty based on the maternal environment. This hypothesis is different from the *thrifty genotype hypothesis,* which states that natural selection for particular genes in some human populations makes these individuals more metabolically efficient. This theory has been implicated in diseases among Native Americans who are now subjected to the modern American diet.

## Little changes mean a lot

Given that environmental factors can alter developmental trajectories (think queen bee), it's not surprising that changes in the DNA sequence of developmental genes will do so as well. In fact, as you probably can guess, small changes in developmental genes can have tremendous impact on the adult organism. Take flies that have grown legs where their eyes should be (see Figure 14-1). This example is a little bit Frankenstein-like (and not too good for the poor fly that happened to have this mutation), but it's a wonderful example of a small change in development that has a big change in morphology.

**Figure 14-1:**
A mutant fly.

What the fly example shows is how small mutational changes in regulatory systems can influence the diversification of multi-cellular organisms. Scientists had been struggling to reconcile how the small mutations we see all the time — a DNA letter or two changed here, a few bases deleted there — could result in major changes in body plan. Now scientists know from laboratory evidence that tiny changes in DNA sequence can have major implications.

With the fly, scientists noticed the mutation and tried to figure out what had happened. What they discovered was that a single mutation in a gene that acts early in flies' larval development results in antennae where eyes should be. In other words, while this fly was still just an innocent little maggot, the pattern of development that resulted in the adult fly had already been determined.

A small change in a single gene involved in development can cause a large change in the final phenotype.

# Key Ideas about Evo-Devo

As I'm fond of mentioning throughout this book, biology has come a long way since Darwin's day, and the field of developmental biology is another good example. Today, scientists are busy trying to figure out exactly how genes work; in Darwin's time, they were struggling just to understand what genes might be. Absent all the molecular techniques available today, it's no surprise that the study of embryology in the mid-1800s involved primarily the examination of embryos from different stages of development.

The following sections look at what early embryologists found — or, more accurately, what they *thought* they found — and explains what scientists know today.

## Developmental stages = Evolutionary stages

According to biologist Ernst Haeckel (1834–1919), the developmental stages that an embryo passes through reveal the evolutionary history of the species. In other words — and brace yourself; it's a mouthful — the *ontogeny* (the developmental process) *recapitulates* (summarizes) *phylogeny* (the evolutionary history).

Haeckel saw in the various stages of mammalian development what he described as stages of development corresponding to specific ancestral species. At one point in their development, for example, human embryos have a tail-like structure, which was thought to signify tailed ancestors. Earlier developmental stages included pharyngeal arches, which were thought to resemble gill slits. This feature was taken to indicate that the developing human embryo passes through a fishlike stage.

Haeckel's investigations were conducted during a period when Darwin's theory of evolution by natural selection and the concept of the common ancestry of all life were stimulating a fair amount of research. The proposition that ontogeny recapitulates phylogeny was seen as being consistent with Darwinian theory, though it was difficult to understand why such a pattern would occur.

Today, scientists understand that the different stages a developing embryo goes through are in no way the equivalents of other species in the tree of life. There is no fish stage in mammalian development, for example. But humans do have structures that are *homologous to* (share the same ancestor with) structures that a fish has, and some of these structures are most evident at early embryological stages.

Two traits that are homologous have a common ancestor. The wings of birds, for example, are all homologous. All birds are descended from an initial bird ancestor. Go back farther, and you can see that the front limbs of all *tetrapods* (four-footed animals and the things descended from them — so whales and snakes are called tetrapods, too, as are humans) are homologous. Your arms, the wings of all birds, the front legs of crocodiles, and the front flippers of dolphins all trace back to the first tetrapod. Homologous characters can be similar in appearance, but they don't have to be. Contrast that with *analogous*. The wings of birds and the wings of butterflies are analogous. They have similar functions, but not as a result of having a common evolutionary ancestor. Insect wings are derived from very different parts than are the front limbs of ducks, for example.

Although Haeckel's idea that ontogeny recapitulates phylogeny turns out not to be true, he was right in thinking that embryology can tell scientists something about evolution. Today, 150 years later, scientists have a much better handle on what that something is.

## Earlier vs. later stages

For related species, earlier stages of development are usually more similar than later stages. In the early stages of human and chimpanzee development, the fetuses are very similar, but as development proceeds, the developmental pathways diverge.

For a more detailed example, take crustaceans. Crustaceans include critters you've heard of, such as crabs, lobsters, and shrimp. But the group also contains some members that aren't immediately recognizable as being related to crabs and lobsters. Perhaps the best example is the barnacle. Barnacles, believe it or not, are really quite like lobsters. Although an adult lobster (mmm, mmm, good) is quite a bit different from an adult barnacle (not delicious at all), the early larval stages are extremely similar.

In the early stages of development, the different members of a group are extremely similar, but as development proceeds, the specific adult features of each group are expressed. From this fact, researchers discovered that much of the diversification in body structure within the crustacean group results from developmental shifts later in development.

## It's all in the timing

Haeckel was a clever guy, and although he didn't quite hit the nail on the head with ontogeny recapitulating phylogeny, he did recognize the importance of developmental timing in differences among species. Developmental timing is also important within the same species, with small changes during the embryonic level having dramatic effects at the adult level.

The fancy word for a change in developmental timing is *heterochrony,* which includes any of the following:

> ✔ A change in the start or end of a developmental stage
>
> ✔ A change in the rate of a process
>
> ✔ The loss of the developmental stage

### Human babies

If you like watching nature shows on TV, you've probably seen films of animals being born. The newborn tiger/wildebeest/whatever shakes itself off, stumbles to its feet, and trots along after Mom. If you've had children of your own, you probably noticed that your newborn didn't do that, even though the newborns of all our closest relatives — chimpanzees, gorillas, and so on — are reasonably mobile from day one. Humans are the branch of the primate family tree in which something changed.

To make a long story short, human babies are born before their heads are finished developing. (If you've had a baby pass through your birth canal, you might realize that a malleable skull made delivery easier; if it still didn't seem like a stroll through the park, remember that evolution often involves compromises!) The human child's head and brain continue to develop and grow after birth. This process of juvenile traits persisting later in development is called *neoteny*.

### The Mexican salamander

In a more-extreme form of neoteny, development doesn't progress past the larval stage. One example is the Mexican salamander *Ambystoma mexicanum*, a species that's currently endangered because it lives in a single Mexican lake that's been heavily affected by human activities.

Salamanders typically progress from an aquatic larval stage to a terrestrial stage, but the Mexican salamander doesn't pass beyond the aquatic stage. The gonads mature, but the rest of the body keeps its larval form. It appears that this form originated through mutation in a thyroid hormone, and individuals can be made to change into a form more typical of adult salamanders if you give them hormone injections. A small mutation results in a dramatic shift in development — in the case of this particular salamander species, resulting in an adult that is entirely aquatic.

These salamanders have a couple of interesting features:

> ✔ **They have enhanced regenerative abilities.** They're able to manufacture replacement body parts to a degree much greater than that of salamanders that reach the terrestrial stage. In fact, if they're forced to metamorphose to the typical adult form via thyroid hormone injections, they lose this regenerative ability. It's the typical "Do I want great power or good looks?" dilemma. Unfortunately, these little critters don't get to decide for themselves. Why? Read on.

✔ **They're not uncommon pets in America.** You'd be surprised at the number of parents who let their kids have creatures with fully mature gonads in a larval body that's able to regenerate larval parts as necessary.

### The Caribbean tree frog

You'd think that mutations that prevent development from progressing to the next stage or that eliminate a stage are not good. Interestingly enough, the resulting mutants sometimes survive. In the case of one Caribbean tree frog, they do very well indeed.

When you think of the developmental stages of frogs, you probably think of tadpoles. Well, the Caribbean tree frog *Eleutherodactylus coqui* is a frog with no tadpoles; its developmental pathway has lost the tadpole stage. Eggs develop directly into very tiny frogs.

As a result of this developmental shift, the species is able to live in areas without bodies of water, which tadpoles require for development. These frogs can live in trees, and they colonize mountainous regions where ponds are rare.

Here's an interesting tidbit about these frogs: They've unfortunately and accidentally been introduced into Hawaii, where they're doing extremely well. Hawaii has so many Caribbean tree frogs, in fact, and their calls are so loud that people are reportedly being kept awake! (People who finally do get to sleep are generally awakened again by the thousands of introduced wild chickens — which, on the island of Kauai, have no natural predators, but that's a story for another day.)

## Why any of this is important

The evolution of development is still a young area of evolutionary biology, but it's already extremely important. Understanding the interaction between evolution and development can help scientists figure out the following:

✔ **How specific developmental processes affect the outcome of natural selection.** When researchers know how processes work, they can understand the kinds of changes expected to result from random mutations.

✔ **How the developmental process itself evolves.** The genes that are responsible for determining body pattern, called the *Hox genes* (see the later section "Genes Responsible for Development: Hox Genes"), are slightly different from one class of organism to another. Hox genes of mammals differ from the Hox genes of insects, indicating that the machinery itself is evolving.

✔ **How the great diversity of animals on Earth could have evolved from a common ancestor.** DNA sequence data has allowed researchers to refine the picture of the tree of life and the details of the branching process, but DNA sequence information by itself does not explain how so many varied body plans can have arisen from a common ancestor. Knowledge of developmental controls, as well as laboratory experiments showing how small changes in developmental genes result in large changes in animal body plan (think multi-headed jellyfish), are giving scientists this understanding.

# Genes Responsible for Development: Hox Genes

Certain genes are responsible for some major aspects of animal development. One of the most important discoveries in developmental biology (or in evolutionary biology, biology in general, and medicine, for that matter) is a set of genes called the *Hox genes*. These genes are responsible for the determination of body pattern — a sort of design plan for items such as where the legs go and where the head should be. Pretty important stuff! In effect, these genes control the process whereby the embryo is divided into segments, and then they determine the specific fates of different segments.

The initial research on Hox genes involved a fruit fly, *Drosophila*. Like other arthropods (invertebrates such as spiders, crustaceans, and insects), the fruit fly is made up of a series of segments, some that are the same and some that are different.

If you know what a millipede looks like, you're familiar with an arthropod with a lot of segments that are all pretty much the same. After the head each segment has two pairs of legs — a very simple design and the fossil record tells us that millipedes are a very ancient group of land animals.

Now consider a fly or other insect. The segments are more differentiated:

✔ **Head segments:** This is where . . . well, where the head is. What can I say?

✔ **Thoracic segments**: The thoracic segment includes the internal organs. It's also where the legs are.

✔ **Abdominal segments:** No legs, but this is where the reproductive structures are.

For an up-close-and-personal look at sections, treat yourself by performing a tasty and very informative dissection of a lobster. Find the head and thoracic segments, and then the repeating segments of the tail/abdomen, which are delectable! Note that seeing the segments in the thorax is easier if you take it apart a little bit — externally the thoracic segments are fused together. Not much *in* the lobster thorax is edible (although some people swear that the liver is a delicacy), but it offers a lot of good biology. Lobsters aren't cheap, of course, so it's important that you get the most for your dollar. Education is priceless!

You may be thinking, "Whoa! How can the mammalian body plan be organized by the same family of genes that organized the arthropod body plan, especially when the human body lacks all those repeating segments?" Consider your backbone. It's a structure of repeating segments (all those vertebrae). And at particular points along your spinal column, you have other structures, such as arms and legs.

## Keeping it in the family

As stated previously, Hox genes are the family of genes that control development of body plan. They determine how different areas of the developing embryo become different body parts: where does the head go, where do the legs go — that sort of thing.

The Hox genes are responsible for body patterning in most animals. Even though different animals look completely different, the underlying genes are clearly related. You have them, a mouse has them, a fly has them. Although humans are extremely different from fruit flies, the genes responsible for body patterning are identifiable in both species. The copies in mammals are different from the ones in flies, but not so different in DNA sequence that scientists can't see their common origin in a distant ancestor.

A fly and a mouse have similar Hox genes, for example, but they don't have exactly the same collection of these genes; the exact sequences are different. Also, Hox genes can become duplicated just as can other genes in the imperfect DNA replication process, so different animals have different numbers of Hox genes. Looking across different animals, scientists see cases where some animals have several copies a very similar Hox genes — evidence of past gene duplications — while others will have a different number of copies of related genes.

The fact that the genes for body patterning would be recognizably similar across such different animals was quite a revelation — and a major breakthrough in the evo-devo field. It gives scientists deep insights into how changes at the level of the DNA can result in changes in animal body plan.

Here's the big take-home message: Developmentally you're not all that different from a fly. Understanding Hox genes takes us a long way toward understanding how small changes at the level of the DNA, besides the things we actually know to happen, could result in large differences in animal form.

## Of mice and men and . . . jellyfish!

The preceding discussion focuses on *bilaterally symmetrical animals* — those with a right and left side, a head, and a tail end. Most animals are bilaterally symmetrical: you, your dog, your goldfish, your hamster, and your houseflies.

But a few animals aren't symmetrical — jellyfish, for example. As it turns out, jellyfish also have genes related to the Hox cluster. They're much different from the genes that humans share with mice or even flies, which makes sense, because the hypothesized common ancestor of the bilateral animals and the jellyfish lived millions and millions (and millions) of years ago.

Now, a jellyfish doesn't have a head in the sense that humans do, but it does have a place where the mouth is, which is probably as close to a head as you can imagine. Jellyfish have an oral side and an *aboral* side — in plain English, a mouth side and a side opposite the mouth side. Genes related to the Hox genes in bilateral animals are responsible for this oral-aboral patterning, and in honor of the group that jellyfish find themselves in — the Cnidarians — these genes are called Cnox, instead of Hox, genes. (The C is silent; it's just there to mess with you.)

Cnox genes are similar to the ones involved in head formation in bilateral animals. Scientists can investigate the function of these genes in jellyfish by altering them genetically. One such experiment resulted in a jellyfish with multiple heads and multiple functional mouths!

# Chapter 15

# Molecular Evolution

- - - - - - - - - - - - - - - - - - - - - - - - - - - - - - - - - - - - - - - - - - - - - - - -

## In This Chapter

▶ Understanding what genomes do (and don't do)

▶ Deciphering genomes

▶ Distinguishing between coding and non-coding DNA

▶ Discovering neutral mutations

▶ Telling time with the molecular clock

- - - - - - - - - - - - - - - - - - - - - - - - - - - - - - - - - - - - - - - - - - - - - - - -

*E*volution is all about heritable changes, and DNA is the material that's inherited. Other chapters in this book don't go into a lot of detail about the various sorts of changes that can occur at the DNA level. Instead, they focus mostly on examples involving changes in alleles at particular loci, like changes in some bit of bacterial machinery that renders the bacteria antibiotic resistant. This simple process, whereby one of the nucleotides (A, C, T, or G) in a bacterium's genome was incorrectly copied and thus changed the bacterium's phenotype, is an example of a change at a locus from one allele to another and an example of how a new allele appears in a population.

But there are other evolutionary questions we can ask about an organism's DNA. For example, how many genes are there? It turns out organisms don't have the same number of genes. Since all organisms share a common ancestor, where did the new genes come from? What sort of evolutionary changes can result in new instructions in the organism's instruction manual? Another question is how much DNA is there, and is all of it genes? Scientists have discovered that the number varies a fair bit between organisms and not always in a way you'd expect. In some organisms, most of the DNA doesn't correspond to different genes, and in other organisms, such as ourselves, lots of the DNA doesn't seem to do very much.

The field of molecular evolution seeks to understand how these changes come about, how evolution works at the DNA level, and what understanding the details of the process can tell us about how evolution might proceed.

# My Genome's Bigger than Your Genome!

As discussed in Chapter 3, every organism has a genome made up of DNA, which is the instruction manual for making the organism, spelled out in a four-letter alphabet: A, C, T, and G. These letters are called *nucleotides,* or bases. The size of the genome is the total number of DNA bases used to spell out the instructions. (OK, not every organism has a genome made up of DNA; some types of viruses have genomes made of RNA instead of DNA. But that exception's not important here.)

## Genome sizes at a glance

Haploid organisms have only one copy of their DNA; diploid organisms have two copies. Humans are diploid: A person's genome consists of two DNA copies, one from Mom and one from Dad. Both copies contain the same type of genes (eye-color genes, for example), whose "specifics" (blue eyes versus brown eyes, for example) may or may not be different.

To standardize across all organisms, when scientists talk about genome size, they talk about the size of a haploid genome. For diploid organisms, genome size corresponds to the amount of DNA in a non-fertilized egg or in a sperm cell. Table 15-1 lists the haploid genome sizes and the number of genes for many organisms. It also identifies what branch of the tree of life the organism occupies.

| Table 15-1 | | Genome Sizes | | |
|---|---|---|---|---|
| *Organism* | *Branch of Tree of Life* | *Estimated Size (Mb= million bases)* | *Estimated Number of Genes* | *Average Gene Density\** |
| Humans | Eukaryote | 2900 Mb | 30,000 | 1 gene per 100,000 bases |
| Mice | Eukaryote | 2500 Mb | 30,000 | 1 gene per 100,000 bases |
| Fruit flies | Eukaryote | 180 Mb | 13,600 | 1 gene per 9,000 bases |
| *Arabidopsis thaliana* (a little flowering plant) | Eukaryote | 125 Mb | 25,500 | 1 gene per 4,000 bases |
| Round worm | Eukaryote | 97 Mb | 19,100 | 1 gene per 5,000 bases |

| Organism | Branch of Tree of Life | Estimated Size (Mb= million bases) | Estimated Number of Genes | Average Gene Density* |
|---|---|---|---|---|
| Yeast | Eukaryote | 12 Mb | 6,300 | 1 gene per 2,000 bases |
| E. coli | Eubacteria | 4.7 Mb | 3,200 | 1 gene per 1,400 bases |
| H. influenzae (can cause blood poisoning and meningitis) | Eubacteria | 1.8 Mb | 1,700 | 1 gene per 1,000 bases |
| Rice | Eukaryote | 430 Mb | 32,000–56,000 | 1 gene per 10,000 bases |
| Entamoeba histolytica (a single-celled amoeba) | Eukaryote | 24 Mb | 10,000 | 1 gene per 4,000 bases |

*Average gene density refers to how many bases of DNA there are for each gene.

*Note:* The reason the numbers in Table 15-1 differ between "Estimated Size" and "Estimated Number of Genes" is that the "Estimated Size" includes both coding and non-coding DNA (explained in the section "Distinguishing between genes and non-coding DNA"), whereas the "Estimated Number of Genes" entries include only the coding DNA.

The tree of life (refer to Chapter 9) has three main branches:

- **The Eubacteria:** These are all the bacteria you've heard of, including *E. coli,* staph, strep, and other such critters

- **The Archea:** These are the other group of single-celled things without a nucleus. These look pretty similar to the Eubacteria under a microscope but turn out to be very different when scientists are able to figure out their DNA sequences.

- **The Eukaryotes:** These include all the organisms whose cells have a nucleus — yeast, pine trees, you, and so on — basically anything big enough to see as well as most of the biggest things that are still too small to see.

## The C value and the C-value paradox

When scientists began to measure the genome sizes of different organisms, two things became apparent: Within a species, genome sizes are the same, but across species they differ quite a bit and not necessarily in the way you'd expect. The following sections explain.

### Genome sizes consistent within a species

Within a species, every organism has the same size of genome. This finding makes perfect sense. The instruction manual to make one person should be the same length as the instruction manual to make somebody else, although the details vary from person to person. Both people need the instructions to make eyes, for example, but the exact details — the color and shape of the eyes — may vary from person to person.

What's true for people is true for other species as well: The instruction manual is the same length, even if some of the instructions are slightly different. There are some exceptions to this, however, such as with E. coli, whose genome size can vary, as explained in the later section, "Getting genes from other lines: Lateral gene transfer."

Because the size of the genome is constant across all individuals in a species, a species' genome size is referred to as its *C value,* with *C* standing for *constant.*

### Genome sizes vary between species

Between species, genome size varies greatly — a fact that is extremely puzzling, because although it makes sense that different organisms require different-size instruction manuals, no obvious connection exists between the size of the species' genome and that species' complexity. For that reason, scientists call the discrepancy between complexity and genome size the *C-value paradox* (or the *C-value enigma*).

## Distinguishing between genes and non-coding DNA

An organism's genome can roughly be divided into two parts:

- ✔ **Genes (coding DNA):** These sequences of DNA are transcribed and are the genes that determine phenotype.

  During transcription, DNA sequences are copied to RNA. During another process called *translation,* the RNA is copied to amino acids, chains of which are called *proteins.* Not all the transcribed RNAs are translated into proteins; they have some other jobs. Chapter 3 has the details on these processes.

✔ **Non-coding DNA.** These areas of DNA aren't transcribed. In other words, they don't seem to do anything.

### Number of genes

If you take a close look at the numbers in Table 15-1, you may notice that some of the differences make sense. You'd probably expect single-celled organisms to have fewer genes than multicelled organisms, and that's what you find in some instances. Humans, for example, have about 20,000 genes, whereas *E. coli,* a species of Eubacteria (beneficial bacteria) that inhabits the human gut, has slightly more than 4,000 genes.

But in other instances, the numbers aren't what you'd expect. Although humans have twice as many genes as fruit flies, rice plants have almost twice as many genes as we do. At first pass, rice isn't obviously twice as complex as humans are. Because scientists don't know what most of the rice genes do, they don't really understand why rice has so many genes, but they do know that having this many genes isn't a universal property of plants. The small weed *Arabidopsis* has about the same number of genes as humans but far fewer than the rice plant has.

While it's true that the littlest critters — viruses, eubacteria, and archea — have the smallest genomes and the smallest numbers of genes, there are other single-celled creatures, like certain amoebas, that have enormous genomes and the same number of genes as some (but not all) multicellular organisms. The single-celled amoeba *Entamoeba histolytica* has almost 10,000 genes, not that many fewer than a fly!

### Amount of non-coding DNA

Another thing you may notice in Table 15-1 is that different organisms have different amounts of non-coding DNA, represented in the "Average Gene Density" column. The more bases there are for each gene indicates more non-coding DNA. So humans, who have only one gene for every 100,000 bases, have quite a bit of "wasted" space, or non-coding DNA. *H. influenzae,* on the other hand, has one gene per thousand bases, meaning it has virtually no non-coding DNA.

Clear patterns appear between the major groups of organisms:

✔ Eubacteria and Archea have almost no non-coding DNA.

✔ Eukaryotes have non-coding DNA, but the amount of non-coding DNA varies widely among them. Some ferns have 100 times as much non-coding DNA as humans do.

✔ Viruses don't have non-coding DNA, but they don't fit neatly on the tree of life. In fact, viruses probably are not a single group. They have such small genomes that very little information is available to group them with other organisms.

No one knows exactly why organisms on the different branches of the tree of life have different amounts of non-coding DNA, although scientists can make educated guesses:

- ✔ **Size of the organism:** It seems reasonable that the smallest organisms simply don't have room for extra stuff. You can't fit a gallon of milk into a quart container. The same constraints wouldn't exist for eukaryotic cells. Your genome has a lot of non-coding DNA, but the nucleus is still only a small part of the cell; it seems to have room to spare.

- ✔ **Rapid cell division:** For organisms such as viruses and bacteria, for which rapid division is a key component of fitness, the extra time that replicating a larger genome takes is too much of a selective disadvantage, so non-coding DNA doesn't accumulate.

- ✔ **Population size:** Maybe non-coding DNA can accumulate more easily in eukaryotic organisms because they have smaller population sizes, on which genetic drift (random events) can be a more influential evolutionary force. (Head to Chapter 6 for info on genetic drift.)

# The Whys and Wherefores of Non-coding DNA

Imagine being the person who took the first gander ever at a genome, the instruction manual for life. Pretty amazing stuff. Now imagine ferreting out which bits do what and discovering that quite a lot of what you're looking at doesn't seem to do anything. It's just there, cluttering things up. In a word, it's junk. Quite a bit of the genomes of eukaryotic organisms is junk — non-coding DNA. Why is it there? No one really knows, but several explanations have been proposed, as the following sections explain.

## It performs a function

Maybe non-coding DNA plays some role in controlling how other DNA sections are transcribed, even though it isn't transcribed itself. In this case, the non-coding DNA is advantageous to the organism (it performs a necessary job), so natural selection maintains it. Some evidence exists that this situation actually occurs; even so, it's not enough to account for the huge amounts of non-coding DNA.

Alternatively, the non-coding DNA may serve a structural function during cell division or the production of *gametes* (sperm and egg). Replicating the eukaryotic genome, which is packaged in a series of chromosomes (refer to Chapter 3 for info on chromosomes), is a pretty complex process. The non-coding DNA could be involved in putting the chromosome together — pairing things up and partitioning the copied chromosome in the daughter cells, for example.

# It serves no function but isn't harmful

Some scientists hypothesize that the non-coding DNA serves no function, but because it's not especially harmful, natural selection doesn't select against it. If a certain amount of non-coding DNA doesn't have any negative fitness costs, it could just pile up in the genome and persist because it keeps getting dragged along to future generations by nearby advantageous genes. (Head to Chapter 4 for more on hitchhiking DNA.)

# It's parasitic!

Much of the non-coding DNA may be parasitic DNA — the result of replicating elements in the eukaryotic genome reproducing themselves. As you can imagine, anything that's a parasite can't be good for the host organism, and selection would act to favor individuals with less of this parasitic DNA.

So why is it still there? Because parasitic DNA elements that are best able to reproduce themselves in the eukaryotic genome are selectively favored, even though selection is also acting to favor organisms in which they are less able to replicate. Lots of evidence exists that a large amount of your genome is really taken up by the selfish elements.

## Retroelements

All kinds of selfish, non-coding DNA are around, but one very important kind is the retroelements. To understand the retroelements, think about retroviruses, the most famous of which is HIV (explained in detail in Chapter 18). Retroviruses are viruses that alternate between an RNA genome and a DNA genome. They start with an RNA genome, infect their host, make a DNA copy to integrate into the host, and then replicate an RNA copy so that they can spread.

A retroelement does essentially the same thing that a retrovirus does, except that it lacks the genes that enable it to package the RNA copy into a particle that spreads. (When scientists sequence retroelements, they find that they are closely related to retroviruses but have a reduced set of genes.)

Amazingly, 95 percent of the human genome is made up of retroelement-like sequences. What's their purpose? Like everything else in evolutionary biology, their purpose is simply to make more of themselves. They replicate in your genome, and you pass them along to your kids.

### *Harmless — until they mess things up*

For the most part, retroelements don't seem to do you any harm — until they do. Retroelements can move around in your genome and mess things up. They can insert into a new location right in the middle of an existing gene.

An example of a retroelement in action is the blue merle pattern in some breeds of dog. This coloring pattern is caused by a particular retroelement popping in and out of the pigment genes during embryonic development. At a certain point, the retroelement stops jumping around, but the dog still ends up with splotches of different colors in different places. Another, more harmful effect of retroelements may be cancer, where certain retroelements move around more actively.

Bottom line: Retroelements don't do anything beneficial, take up space, can be harmful, and may be parasitic. You'd be better off if they weren't there, but they just keep reproducing.

Transposable elements in general (pieces of DNA which pop around from one place to the next in the genome) and retroelements in particular (a subclass of transposable elements whose replication involves a reverse transcription step — that is a transcription from RNA back to DNA) are a big part of your genome, but they don't do anything of value. Why oh why are they there? That's the sixty four million dollar question of genome size.

# Coding DNA: Changing the Number of Genes an Organism Has

Although the largest differences in the amount of DNA between different species are the result non-coding DNA (see the preceding section), differences also occur in the number of genes between different organisms.

Think back to the branching tree of life (refer to Chapter 9) and the evolution of more complex organisms such as humans from less-complex organisms. You can appreciate why the number of genes has increased. In this section, I explain how the number of genes an organism has can change. New genes can appear in a variety of ways.

## Getting genes from other lines: Lateral gene transfer

*Lateral gene transfer* is the process by which one evolving lineage acquires genes from other lineages. This mechanism isn't thought to be especially important for eukaryotes such as humans, but it's quite common among non-eukaryotes, the Eubacteria and Archea.

A good example of lateral gene transfer is the vastly different number of genes between the beneficial *E. coli* strain that all humans have in their guts and the disease-causing *E. coli* 0157:H7. The human-gut *E. coli* has about 4,300 genes, whereas the pathogenic strain has 5,400 — a huge difference. It's not clear where all these extra genes came from, but some of them clearly are related to genes from the bacteria species *Shigella dysenteriae,* whose name probably makes clear why these genes can turn a good *E. coli* bad.

## Shuffling exons: Alternative gene splicing

DNA sequences are transcribed into RNA sequences. A subset of these RNA sequences, termed *messenger RNA* (or mRNA for short), are then translated into sequences of amino acids, called proteins.

The most sensible way for an organism to make proteins would be to have just enough DNA to code for the length and type of protein being made. And often that's what we find, especially in things like bacteria. But many times, the sequence of DNA, and thus mRNA, is much longer than necessary to make the desired protein.

In this too-long sequence, some sections of the RNA nucleotides are translated to amino acids (they're called *exons*), and others aren't (they're called *introns*). To make the amino acid, the mRNA is "processed" — the introns get spliced out, leaving just the exons all strung together. This new, shorter piece of mRNA codes for the amino acids that make the protein, and everything's hunky-dory.

Yet when scientists look at the details of this process, they find that sometimes there's more that one way to process the same mRNA.

- ✔ Sometimes all the exons get strung together
- ✔ Other times just some of the exons get strung together.

In the second case, when the exons are sewn back together to make the final piece of RNA, some of them get left out, and the result is that different proteins are made from a single gene. This is important, because using different combinations of exons from the same sequence of DNA can result in cells with different functions. This process is called *alternative splicing* or, more informally, *exon shuffling*.

In exon shuffling, a gene with four exons, for example, might be spliced differently to create several different types of mRNA. One obvious one would be an mRNA made up of all 4 exons. But mRNAs could also be made from just a subset of the exons — say exons 1, 2, and 4 in one case, and exons 1, 3, and 4 in another. In each of these cases, the protein produced from this mRNA could have a different function. In mammals, for example, the calcitonin gene produces a hormone in one cell type and a neurotransmitter in another cell type, due to alternative splicing.

Alternative splicing suggests one way that new functions can arise. A mutation that resulted in exons being spliced one way sometime and another way another time would create two protein products from the same DNA. In short, through exon shuffling, it would be possible to gain a new protein while still being able to make the original one. If the new protein were selectively advantageous, then the new mutant would increase in future generations.

## Duplicating genes: A gene is born

Errors in DNA replication can lead to duplications of sections of DNA or, as in the creation of polyploids, duplication of the entire genome (see Chapter 4). It's akin to going to Kinko's for a single copy and ending up with two: the copy you needed and an extra you're not quite sure what to do with. The same thing goes for the DNA. What's an extra gene good for?

If a mutation knocks out a duplicate copy, the organism is pretty much back where it started: It still has the original copy and an extra. This situation is the spare-tire gene duplication theory. (Just kidding.) The extra copy may or may not be functional:

- ✔ **When the second copy is nonfunctional:** These nonfunctional genes are called *pseudo-genes*. Pseudo-genes look like the original genes, but they're a little, or a lot, broken, and they tend to accumulate more and more mutations over time. The more mutations occur, the harder it is to recognize that these genes are related to existing genes. But since they're already nonfunctional, there's no fitness cost to a few more mutations.

- ✔ **When the second copy is functional:** If the second copy mutates such that it is able to perform an additional function that's selectively advantageous, individuals with this mutation increase in frequency, resulting in two different yet related functional genes where one originally was. Many examples exist of families of genes that appear to have a common ancestor.

When you know about the possibility of gene duplications, you understand that natural selection could result in a change of function of one copy without having to worry about losing the function of the other copy. When you have two copies, you have room to move. Because most mutations are deleterious, these duplicated genes usually end up losing their original function without acquiring any new function. But sometimes beneficial mutations occur in the second copy, and a new gene is born.

# The Neutral Theory of Molecular Evolution

The neutral theory of molecular evolution says that genetic drift — random events that affect evolutionary change (see Chapter 6) — accounts for much of the change in DNA. This is the case because most mutations are selectively neutral. In fact, much more variation is neutral than scientists once thought. The variation exists in the DNA, but either it doesn't result in a change at the level of the protein or, if it does result in changes in proteins, these changes don't change the protein's function.

The chance that a mutation will have no effect can vary between different genes. Some proteins, for example, are very tightly constrained in the shapes that they can adopt and still be functional. For these proteins, relatively few mutations are selectively neutral. When scientists examine the mutation rates of different proteins within the same organism, they find that some evolve faster than others.

As stated previously, neutral mutations are neither good nor bad, and when a mutation is neutral, natural selection doesn't act on it. (Why should it? A neutral mutation doesn't help the organism, which would cause an increase in frequency in subsequent generations. Neither does it hurt the organism, which would cause a decrease in frequency.) Therefore, the evolutionary force that acts on these genes is genetic drift. Over enough time, a selectively neutral mutation can reach a frequency of 100 percent in a population just by chance — a situation called *fixation*. (You can read more about fixation and genetic drift in Chapter 6.)

Many mutations are almost, but not quite, neutral. A slightly deleterious mutation, for example, might still increase in frequency as a result of genetic drift in a small population. Remember that for any given mutation, the chance of fixation (that is, the chance of reaching a frequency of 100 percent) is a function of population size. If population size fluctuates (as it often does), a particular gene may be changing in frequency primarily as a result of natural selection at one time and primarily due to genetic drift at another time.

Two evolutionary forces are at work: natural selection and genetic drift. If a mutation isn't neutral, both natural selection and genetic drift can be the cause of evolution. If a mutation *is* neutral, only genetic drift can result in a change in the frequency of the gene over time. (This discovery — that random events are evolutionary forces in and of themselves — has been the most important addition to the theory of evolution since Darwin.)

# Telling Time with Genes: The Molecular Clock

The *molecular clock* refers to the idea that if mutations are often neutral, all other things being equal, they might be expected to accumulate at a constant rate. If this idea is true, it should be possible to measure the genetic differences between two species and determine how long ago the two lineages diverged. Sounds, good, but telling time with the molecular clock is tricky; sometimes you can, and sometimes you can't.

## When you can't

Using the molecular clock to determine when lineages diverged requires that the neutral mutations accumulate at a constant rate. Which doesn't happen, for several reasons:

- **Differences between genes:** There is no reason to think that the proportion of neutral mutations should be constant across genes with different functions. One protein may function only if it's exactly the right shape or configuration to do its job; another protein may do its job pretty well even in spite of a few changes. At the very least, comparisons should be made only between the same gene in different species. But even then, scientists can't be sure that a gene shared across two species performs the same function. The function of the gene may have changed as the species evolved to a different environment, for example.

- **Population size and generation time:** The rate at which neutral mutations become fixed in a population is a function of population size and generation time, both of which may vary between lineages after they diverge. In small populations with short generation times, neutral mutations rise to fixation more rapidly.

- **Strength of selection:** The strength of natural selection may vary between species over time, thus changing the ratio of neutral to non-neutral mutations that rise to fixation.

## When you can

Despite all the reasons why you shouldn't get too excited about the idea of the molecular clock (see the preceding section), in some cases, scientists can show that the accumulation of mutations is an excellent predictor of the time since two species diverged.

The key to using the molecular clock effectively is calibrating it. To do that, you must know the time at which some lineages diverged so that you can translate the amount of divergence between different lineages into time. Then you can use this information to estimate the divergence times of additional species.

## Performing experiments in the lab

For organisms that have generation times short enough to create divergent lineages in the laboratory, you can ask how fast mutations accumulate in these microorganisms over time. Then you can look at sequence variation that you know to be neutral — changes in the DNA that don't result in any changes in the protein. These changes pile up over time, based on the mutation rate. You measure how fast they pile up, and then you know the mutation rate — simple as that! (This strategy doesn't work for organisms that are much longer lived, of course.)

## Looking at ancient DNA and the fossil record

As researchers' biochemical techniques become more sophisticated, they've been able to retrieve DNA sequences from the distant (but quantifiable) past. By studying these ancient sequences, they've been able to put bounds on the rates of mutation accumulation between the date of the old sequence and the current time.

The fossil record also allows scientists to generate estimates of when different lineages diverged by noting the geological era in which the fossil was found. Then they can use these divergence times to calculate the rate at which mutations have accumulated in the lineages since the divergence.

Imagine you have three species: species 1, species 2, and species 3. You sequence all three, generate a phylogeny, and find that species 2 and 3 seem most closely related, and they're both related to species 1. Just by digging and finding fossils at different ages, you have a pretty good idea of when species 1 and 2 split. But you have absolutely no fossils telling you anything about the history of species 3, and you want know when species 3 split off from species 2.

By knowing when species 1 and 2 diverged, you can correlate the number of DNA differences between these two species and the amount of time since they diverged. You can then take this estimate and use it to translate the number of DNA differences between species 2 and 3 into the time since they diverged, even in the absence of any fossil record for species 3. And that's the molecular clock!

From the fossil record, scientists know the approximate times when many lineages arose. Date the rocks that a fossil is found in, and you pretty effectively date the fossil. Scientists know, for example, about when the mammal lineage split off from the rest of the tetrapods (four-limbed creatures, such as lizards, turtles, and birds). On a finer scale, they have a pretty good idea about when the hominid lineage diverged from the chimpanzee lineage.

Scientists can use the fossil record to generate divergence times for lineages for which there is a good fossil record, and then they can use the molecular clock to estimate divergence times for species for which a good fossil record isn't available.

### Examining biogeographic patterns

Biogeographic patterns can also generate estimates of divergence times. Take, for example, the fruit fly species *(Drosophila)* that lives in the Hawaiian islands. Because geologists have an excellent understanding of how the Hawaiian islands formed, they can date the islands accurately.

The Earth's crust has a thin spot, and as the Pacific plate moves across this hot spot, periodic eruptions have generated the chain of the Hawaiian islands. Because geologists can date exactly the age at which lava solidifies, they can figure out when the islands were formed.

In addition, the Hawaiian islands are extremely distant from other land masses. As a result, much of the biological diversity on the islands is a result of speciation events that happened in Hawaii. Most of the fruit fly species in Hawaii occurs nowhere else, for example. Hawaii is so far from anywhere else that the rate at which fruit fly speciation occurs on the islands far exceeds the rate at which non-Hawaiian fruit fly species could arrive. So although some fruit flies got to Hawaii initially from elsewhere, the original colonists have radiated into many of the species you find there today.

As new islands appear in the chain, flies from the neighboring island colonize them and, over time, diverge to become separate species. As explained in Chapter 8, this divergence is a consequence of the very reduced rates of genetic exchange between the islands. A fruit fly occasionally gets from one island to another, but this migration doesn't happen often enough to overwhelm the gradual divergence between the separated populations and their subsequent speciation.

Phylogenetic analysis (refer to Chapter 9) of the Hawaiian fruit fly yields a tree that matches geologists' understanding of the geological formation of the islands — and sure enough, the fly species on different islands each share a most recent common ancestor with the species on the next island over.

Usually, we don't know when in the past two lineages became geographically separated, but with the Hawaiian islands, we know exactly when the new islands popped up out of the ocean. We can combine the data about how genetically different two species are (which we get from the sequence of their DNA) with the length of time they've been separate species.

Not surprisingly, the longer two species have been separated, the more genetically different they will be, simply because changes add up over time. But what's most important about these Hawaiian flies is that, because we know the dates the islands appeared, we can tell that the amount of genetic difference is exactly correlated with the length of time since the species diverged. If one pair of species diverged twice as long ago as another pair, it has twice the amount of genetic differences.

This information tells us that the molecular clock can tick at a constant rate for long periods of time. Proving that (through studying the phylogeny of Hawaiian fruit flies and knowing the dates when the islands formed) makes us more comfortable with assuming that the rate of molecular evolution may be constant in other species as well.

# Part IV
# Evolution and Your World

"Ooo! This would make a wonderful souvenir."

# In this part . . .

Although many of us like to tell ourselves that we're different from other organisms in some fundamental ways, we're all subject to evolutionary processes. So evolution doesn't pertain only to animals, plants, bacteria, etc. It pertains to humans as well. As this part explains, we have our own evolutionary history as a species and an evolutionary future, as well.

Ironically, the other species that poses the biggest challenge — and threat — to us and therefore deserves attention in this part is the one we can't even see (without help): the microbes, like bacteria and viruses, organisms that evolve amazingly quickly and in response to the medicines we use to fight them.

# Chapter 16

# Human Evolution

........................................................

........................................................

Most people are very interested in human evolution and know something about it but still find the topic perplexing. They have an inkling, for example, that humans evolved from apes (actually, apelike creatures) but think that evolution works differently for humans than it does for other organisms. It doesn't. The same evolutionary principles that apply to every living organism apply to human beings: speciation, genetic drift, coalescence, you name it.

This chapter explains both the evolutionary origin of our species and subsequent evolution within the species *Homo sapiens*. In a way, this chapter answers two questions: "Where did we come from?" and "Where are we going?"

## The Origin of Homo Sapiens: Where We Came From

When you think about human evolution, you may think immediately about the fossil record. Although fossils are extremely important parts of the evidence we humans have for understanding our own origins, other vital lines of evidence exist as well:

 ✔ **Phylogenetic reconstructions:** A *phylogenetic reconstruction* is essentially a visual representation of the genealogy of a group of species (refer to Chapter 9 for more details). This image can provide insight into where humans fit in the tree of life.

✔ **Human DNA studies:** By looking at our DNA, researchers can get intriguing information about the patterns of migration of *Homo sapiens* (humans) that help them sort out different hypotheses about where humans originated. They can't get this info just by looking at fossils, which aren't always clear.

✔ **Neanderthal DNA studies:** It's possible now to obtain DNA from one of our closest relatives: the Neanderthals. By comparing modern human DNA with Neanderthal DNA, scientists gain a better understanding of our relationship to this extinct species of hominid.

## Phylogenetic evidence: Hangin' round on the Tree of Life

As I explain in Chapter 9, phylogenetics takes data about existing species and reconstructs the evolutionary branching pattern that led to those species. Not surprisingly, no small amount of effort has been devoted to reconstructing the parts of the tree of life where humans reside. Our particular branch includes the apes: gibbons, orangutans, gorillas, the two types of chimpanzees (the standard one that you're familiar with and the bonobo, which used to be called the pygmy chimp but which turns out to be a species of its own), and us!

When biologists started wondering where all these creatures should reside on our branch of the tree, they imagined one sub-branch leading to us and another sub-branch leading to all those charming, furry creatures who seem so different from us. Researchers could tell that, of all the animals, humans are most like apes, but in the past, they tended to think of the apes as belonging on the other twig of our shared branch of the tree.

Enter the amazing resource of DNA. Now that scientists have been able to sequence human DNA as well as good samples of DNA from the other apes (in the case of the chimpanzees, the entire genome), they've discovered that humans and chimps aren't very distant at all. The current best hypothesis about the relationships between humans and the rest of the great apes is shown in Figure 16-1.

In Figure 16-1, you can see that the two chimpanzee species (chimps and bonobos) have a most recent common ancestor. These species are a lot more like each other than they are like anything else. But — surprise! — humans have a most common ancestor with the chimpanzee lineage that we don't share with the other great apes. To find the most common ancestor of gorillas, humans, and chimpanzees, you have to go back a bit farther — and farther yet to find the most recent common ancestor of orangutans, gorillas, humans, and chimpanzees.

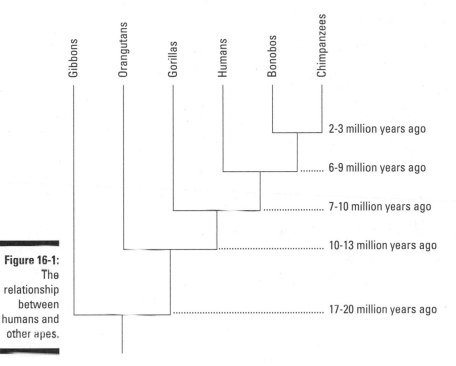

Figure 16-1 shows a point that I want to drive home: *Humans didn't evolve from chimps!* Instead, chimps and humans arose from the same common, apelike ancestor.

### Kissing cousins

Humans turn out to be a lot more similar to chimpanzees than biologists first thought. Human DNA is about 95 percent the same as chimpanzee DNA. A lot of active research is going on in this area, and some studies put the number at 97 percent or 98 percent. Whatever the precise percentage, however, the bottom line is that humans are very similar to chimps.

Why do the different percentages exist? Because it's no easy task to figure out which bits in the DNA sequence are genes, which bits *might* be genes, and which bits are just junk. (Yes, believe it or not, humans, chimps, all other mammals, and most multicellular creatures have a lot of junk DNA; refer to Chapter 15 to find out more about non-coding DNA.) As a consequence, different researchers come up with different estimates regarding which parts are genes and which parts aren't, and thus different estimates of similarity between species.

By combining the information about the relative differences in the genomes of chimpanzees and humans with what we know about the rate of DNA substitutions in specific genes in these two organisms, researchers can tell that, approximately 5 million to 7 million years ago, the lineage leading to modern humans split from the lineage leading to modern chimps. (For information on how to use DNA as a molecular clock to determine the time in the past when two lineages split from a common ancestor, head to Chapter 15.)

If the DNA sequence of humans is so close to that of the chimpanzee, why do the two species look so different? They certainly don't seem to be only 5 percent different (at least, that's what we humans like to think!). The answer is that small changes in regions of the DNA that have a regulatory function can have major effects, as explained in Chapter 14, including examples of genes that may be important in the different developmental trajectories of chimps and humans.

### You say hominid; I say hominian

Biologists commonly name every group; everything that has a common ancestor gets a name. And because humans (the species *Homo sapiens*) are the ones who do the naming, we've made sure that every higher group that includes us also starts with the letters *hom* — *hominid; hominine; hominin* (yes, it means something different from *hominine*); *hominian;* and, of course, *human.*

These names come up a lot in the published studies of human evolution, and they certainly do sound awfully scientific. My opinion? They're nothing but trouble. So in this book, I don't use any of these terms except *hominid,* which refers to all the creatures on the branch of life starting at the common ancestor of chimps and humans, and leading up in time along the human branch. If you need a more precise definition, try this one on for size:

> HOMINID: Modern humans and their extinct relatives, going back to the most common recent ancestor with the chimpanzee lineage.

## Carved in stone: The fossils

As you can probably guess from the excitement with which paleontologists greet each new fossil find, hominid fossils are few and far between. This fact isn't especially surprising, given that great apes tend to have relatively low population densities (and our earliest ancestors probably did, too). That scientists have been able to find fossils of many early hominids at all is a cork-popping event.

### The human species found in fossils

Paleontologists have found fossils for a large number of hominid species, including prehuman primates and human primates. In fact, various hominid species have been identified from fossil remains, as Table 16-1 shows.

*Note:* The *A* or *H* in the species' names is scientific shorthand. Instead of writing *Australopithecus,* for example, scientists simply write *A.* The term *Australopithecus* speaks to the origin of the fossil: southern Africa. *H,* of course, stands for *Homo,* which means *wise.* The name *Homo sapiens* means *wise man.*

| Table 16-1 | Hominid Species |
|---|---|
| *Name* | *Years on Earth (Based on Fossil Finds)* |
| A. anamensis | 4.2 to 3.9 million years ago |
| A. afarensis (Lucy) | 3.6 to 2.9 million years ago |
| A. africanus | 3 to 2 million years ago |
| A. aethiopicus | 2.7 to 2.3 million years ago |
| A. boisei | 2.3 to 1.4 million years ago |
| A. robustus | 1.8 to 1.5 million years ago |
| H. rudolphensis | 2.4 to 1.8 million years ago |
| H. habilis | 2.3 to 1.6 million years ago |
| H. ergaster | 1.9 to 1.4 million years ago |
| H. erectus | 1.9 to 0.3 million years ago (and possible 50,000 years ago) |
| H. heielbergensis | 600,000 to 100,000 years ago |
| H. neanderthalensis | 250,000 to 30,000 years ago |
| H. sapiens | 100,000 years ago to today |

As you can see, for most of the past 3 million or so years, multiple species of hominids have existed at any one time — a situation that persisted until about 25,000 years ago. In short, in the not-too-distant past, we shared Earth with human species other than our own.

The dates in Table 16-1 indicate the years from which fossil specimens of each species have been recovered. (For a review of how scientists determine the dates of fossils, see Chapter 2.) But — and this point is important — the lack of fossil evidence doesn't necessarily mean that a species wasn't around longer than the time periods indicated, only that scientists haven't found it. So even though the current dates indicate that a H. habilis, for example, lived from 2.3 to 1.6 million years ago, if H. habilis fossils dating from 1 million years ago are found, the dates would change. (It's also why you see the range for H. erectus and a note that this species also may have existed up to 50,000 years ago.)

### *The tricky task of separating one species from another*

As Chapter 8 explains, a very useful way for determining whether two individuals are of the same or different species is to determine whether they can interbreed. If the answer is yes, the individuals are of the same species; a no answer means, they are different species. Obviously, this information isn't available for any of our fossil ancestors, so in this case, species names are simply a function of *morphology,* or body structure. All the fossils that look the same are assigned to the same species.

This arrangement sounds easy enough until you consider that the fossil record is sparse and fossil remains of hominids usually are extremely incomplete —part of the leg bone here, half a jaw there, part of the cranium somewhere else. Complicating matters even more is the expectation of finding both juvenile and adult individuals, as well as males and females which may vary in size. As a result, there's a fair bit of argument about whether a new find should be considered a new species or merely another representative of an existing species.

Using skull size and shape, jaw muscles, and limb length can help researchers distinguish between one hominid species and another, as follows:

- **Ratio of forelimb to hind-limb length:** Humans' arms are proportionally much shorter than the arms of chimpanzees, and scientists find different fossil hominids with different arm-length ratios.

   As *bipedal locomotion* (walking upright) developed, forelimb length shortened. (If you want to get persnickety, as scientists tend to do, you can say that forelimbs aren't really arms until the organism is walking on its hind legs — hence, the use of the term *forelimbs.*) Forelimb length is a good way to evaluate which group a fossil belongs to because it's relatively constant for individuals of different ages. The absolute lengths of forelimbs and hind-limbs change as the individual grows, but the ratio of the lengths is consistent over a range of individual sizes.

- **Skull shape:** Humans have proportionally much larger brains than do the other apes.

## This just in

Don't you just love it when science keeps discovering things? Every time anthropologists dig up a new skull, humans' view of the fossil record and what it says about human evolution can change. After a recent bit of digging, researchers now know that *Homo habilis* and *Homo erectus* lived at the same time.

This bit of info changes nothing in the chapter; it just reconfirms that in the past, more than one type of hominid was around at the same time. But it does provide a clearer picture of whether one hominid gave rise to another (as opposed to both having descended from a common ancestor). Some paleontologists thought that perhaps *H. erectus* evolved from *H. habilis*. The new fossil find shows that both species were around together, meaning that (1) they both evolved from something else or (2) *H. erectus* could still have evolved from an offshoot of *H. habilis* (you know — some *habilis* got lost on the way to the office, ended up in a place with different selective pressures, and so on).

✔ **The form and arrangement of teeth (*dentition*):** Humans have markedly different dentition from other apes. We do a lot of our food processing with our hands rather than with our teeth, and we lack the powerful jaw muscles and large teeth that are characteristic of the rest of the primates (a group including the apes and other things like monkeys).

We humans have forelimbs, which just happen to be arms, but because we don't need them to reach to the ground when we walk, they can be shorter. Chimps can walk around on hind legs for a little bit, but soon resort to using their (longer) forelimbs again. Watch a chimp walk next time you're at the zoo; then try it with your comparatively short, little arms — you'll fall on your head! Why the difference? Because humans started to adopt a more upright posture that was selectively favored. The move to upright posture and the decrease in the importance of front limbs for locomotion would have happened at the same time.

## Reconstructing the history of hominid evolution

With all these hominid species, scientists are still trying to figure out which species may have given rise to which other species along the lineage that led to humans. Because no one can say with certainty that a fossil represents a common ancestor of two other species (the fossil may represent a closely related dead end, for example), this task is fairly challenging.

Figure 16-2 shows a hypothetical evolutionary pathway from an ancestral species similar to the extant apes and leading to modern humans. It includes a sequence of intermediate species with increasingly large brains, reduced jaw musculature and dentition, and the evolution of *bipedalism* (walking on just two feet). I've thrown in stone tools and fire just for fun.

The actual intermediate species may not be the ones scientists have already found, or they may be; knowing for sure is impossible. What scientists *do* know is that we can reconstruct the evolutionary events that led from an ape-like ancestor to a modern human through the series of fossil species that have already been found. The following sections describe some of the major players; consider this a sort of hits list of the hominid fossil record.

### Lucy in the sky with diamonds: *Australopithecus afarensis*

This famous fossil, called *Australopithecus afarensis* (*A. afarensis,* for short) is commonly known as Lucy. Named for Lucy in the Beatles song "Lucy in the Sky with Diamonds," which was playing at the time, Lucy was such an important find because a large part of her skeleton was found together, which gave paleontologists a fair bit of confidence in describing the species.

Lucy was found in 1974, but her species lived from about 2.9 million to 3.6 million years ago. She was about 3½ feet tall and weighed 60 pounds. She was bipedal. At the time of her discovery, Lucy was the earliest bipedal hominid that had been discovered. The relative length of her forelimbs is intermediate between that of apes and people.

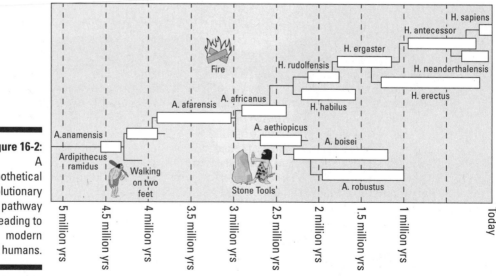

**Figure 16-2:**
A hypothetical evolutionary pathway leading to modern humans.

Before Lucy's discovery, many researchers believed that the driving force behind the evolution of the large brain in the human lineage was that once the hands were free (due to the development of bipedalism) to use the tools, smarter individuals (those who were better toolmakers) would have an advantage. This idea is interesting, but it turns out that Lucy's brain was no bigger than a chimpanzee's — about one third the size of a human brain.

What this fossil species (and others since) have made clear is that the evolution of bipedal locomotion occurred _before_ the evolution of a large brain. A long part of the human family tree is populated with ancient hominids that walked upright but had small brains. This is a nice example of a case in which the evidence provided by a fossil find allows scientists to reject one potential hypothesis about the pathway of human evolution. So it's back to the drawing board to come up with hypotheses that explain why a big brain was all of a sudden favored by natural selection.

The fossil record shows that after _A. afarensis,_ the hominid lineage split into two branches. One branch eventually led to humans; the other branch led to a group referred to as the _robust Australopithecines: A. boisei_ and _A. robustus._ These species had very strong jaws (perhaps for eating plant material). This lineage, which also includes _A. aethiopicus,_ persisted from about 2.7 million to 1 million years ago and then became extinct.

### I'm a traveling man: Homo erectus

Another important branching event in the hominid family tree is the one that separated _H. erectus_ from the lineage leading to _H. sapiens. H. erectus_ originated around 1.9 million years ago and went extinct almost everywhere 300,000 years ago, though one subspecies may have persisted on the island of Java perhaps as recently as 50,000 years ago.

_H. erectus_ had a substantially larger brain than the _Australopithecines,_ made and used tools, and may have been able to control fire. But the species' greatest claim to fame was being the first hominid species to leave Africa, which it did around 1.5 million years ago. By 1.2 million years ago, it had reached China and Southeast Asia.

_H. erectus_ had a brain about two thirds the size of the human brain, but analysis of the internal structure of the _H. erectus_ skull suggests that the area of the brain involved with speech wasn't developed to the extent that it is in later hominids. Meaning? Language came later.

### *Homo sapiens and Homo neanderthalensis*

The last two species on the hominid tree are *Homo sapiens* and *Homo neanderthalensis*. Both species originated in Africa — *H. neanderthalensis* about 250,000 years ago and *H. sapiens* about 100,000 years ago — and both moved out of Africa. *H. neanderthalensis* colonized Europe and parts of Central Asia, whereas *H. sapiens* went on to colonize the whole world. *H. neanderthalensis* coexisted with *H. sapiens* for a long time and went extinct only about 30,000 years ago, but anthropologists don't know why.

*H. neanderthalensis* was more robustly built than *H. sapiens,* possibly as an adaptation to the cold, but the two species' brains were the same size. Evidence has been found that *H. neanderthalensis* had advanced cultures: They modified their environment for shelter; they had art; and they buried their dead (the reason why a good fossil record for the species exists).

Many drawings of *H. neanderthalensis* incorrectly portray the species as having a hunched posture. The reason for the mistake? An early specimen had extreme arthritis. Analysis of many additional specimens reveals an upright posture like that of humans. If you were to pass a specimen of *H. neanderthalensis* on the sidewalk, you might notice the stockier build and facial features (such as a more pronounced brow and perhaps a larger nose), but if he was wearing a nicely tailored suit, you might not give him more than a second glance.

## Hobbit Man: Homo floresienses

An incredible and still somewhat controversial fossil find, *Homo floresienses,* discovered on the Indonesian island of Flores in 2004, was possibly a dwarf species of the genus *Homo.* Nicknamed "Hobbit Man" after the character in J. R. R. Tolkien's *The Lord of the Rings,* the first skeleton found was a female approximately 3 feet tall, weighing perhaps 50 pounds, and estimated to be approximately 30 years old at the time of her death. The fossil evidence suggests that the species inhabited the island as recently as 13,000 years ago. (*Homo Neanderthalensis* went extinct approximately 25,000 years ago.) Because of the limited fossil record (so far, only one fossil has been found with an intact skull), no one knows exactly where *H. floresienses* fits in, but here are some suggestions:

✔ It's a dwarf form of *Homo erectus.*

✔ It's a *Homo sapiens* afflicted with *microcephaly,* a condition characterized by an unusually small head and mental impairments (This idea gained traction based on the fact that the *H. floresienses'* brain case is so small).

✔ It's a new species that lived into modern times — an interesting conjecture that's not supported by any physical fossil evidence but that has traction because people native to the island have legends of small furry people who lived in caves and had a different language. Sightings of such creatures were mentioned as recently as the 1800s and continue to this day on the island of Sumatra.

## Did H. sapiens and H. neanderthalensis interbreed?

The fossil record seems to indicate that *Homo sapiens* and *Homo neanderthalensis* had far more similarities than differences. That being the case, was *H. neanderthalensis* truly a separate species? Maybe, as some researchers have suggested, it was simply another subspecies of *H. sapiens,* and possibly — just possibly — during the many years that the species overlapped in range, they interbred. This possibility is interesting, but you can't find clues in the fossil record. Instead, you have to turn to DNA.

Luckily, DNA is tough stuff, and advanced retrieval techniques have made it possible to obtain sequences of nuclear DNA from *H. neanderthalensis* specimens almost 40,000 years old.

The human and neanderthal DNA sequences are very similar, but there's about one half of a percent (0.5) difference. This is enough to identify specific DNA sequences that are specific to one or the other species.

Here's what DNA testing has revealed: The *H. neanderthalensis* genome is clearly distinct. Had any significant amount of mixing occurred, this result would be less clear. So DNA evidence clearly weighs in against the idea that the two groups interbred. Scientists can't rule out the possibility of any mating between the two species, but they *can* rule out the possibility that any significant genetic mixing occurred.

# Out of Africa: Hominid migration patterns

Human beings originated in Africa. The evidence: Most hominid fossils are found only in Africa, and for those species with a wider distribution, the oldest specimens are always found in Africa. In addition, humans' most closely related living relative (both genetically and as placed on the tree of life; refer to Figure 16-1), the chimpanzee, lives in Africa. Not enough to convince you? The existing genetic variation in the human population provides another line of evidence.

### Coalescence: Sharing a single ancestor

According to the concept of *coalescence,* all the genes in a given population have a single common ancestor — some individual in the past from whom they are all descended. Coalescence is the result of random processes whereby some individuals leave descendents and others don't (refer to Chapter 6).

Think of families in which different members have had different numbers of children. Perhaps one sibling has no children, and another sibling has many. In this way, random forces pile up through time. After enough time has passed, with some members having children and others not, eventually all the existing individuals will be descended from one ancestor. Scientists have discovered that this is essentially what happened, and they dubbed this common ancestor, because she is the one woman from whom all humans descended, "Mitochondrial Eve."

Coalescence doesn't mean that if you go back far enough, you find only one individual who started the whole population ball rolling. What it means is that only one individual, out of however many existed initially, has any descendents left.

Figure 16-3 shows a graphical representation of this process. Time progresses horizontally, from left to right. At the left are eight initial lineages, but as you go forward in time, more and more of these lineages die out due to random events until, in the end, all the current individuals are descendents of just one of the initial eight. Looking backward in time, you can see how the final lineages coalesce to a single ancestor in the past.

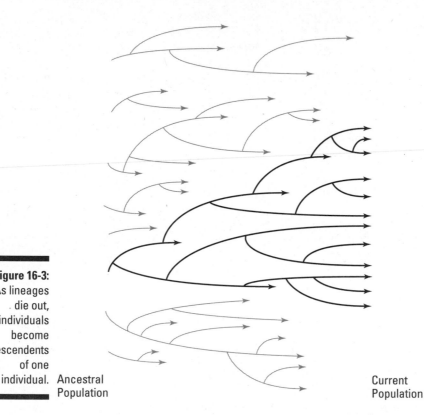

**Figure 16-3:**
As lineages die out, individuals become descendents of one individual.

Ancestral Population

Current Population

We're all descended from Mitochondrial Eve, but we're not all the same. Plenty of variation has been added to the population by mutation over the long period of time between the common ancestor and us. Eve's family tree has lots of branches, and there have been lots of mutation, selection, and genetic drift acting all that time. (Head to Part II for info on the key evolutionary processes.)

### First Africa, then the world!

By sequencing information from the mitochondrial DNA and the Y-chromosomes (see the nearby sidebar "Of mitochondrial DNA and Y-chromosomes" for details on why these bits were used), scientists know the following:

- ✔ Humans probably originated in and spread through Africa first, with a small group leaving Africa and colonizing the rest of the world, although some scientists disagree about whether a single group left Africa and colonized the rest of the world or whether two different periods of emigration from Africa occurred. Figure 16-4 shows a map of the Earth with arrows indicating the patterns of human migration.

- ✔ The mitochondria of all living humans descend from one woman who lived in Africa perhaps 200,000 years ago. Because this one woman has been named "Mitochondrial Eve," many people misunderstand the findings, thinking that the name means that only one woman existed in the beginning — essentially, the Biblical Eve. What the name really means, however, is that only one out of however many women were alive at that time gave rise to all living humans. (This situation is the key point of coalescence, as explained in the preceding section.)

**Figure 16-4:**
Patterns of human migration.

# Of mitochondrial DNA and Y-chromosomes

The name "Mitochondrial Eve" was never meant to be provocative, but unfortunately, most people focus on the "Eve" part (which is merely an allusion to the biblical Eve and *not* meant to imply that there was only a single woman on the planet) and completely skip over the "Mitochondrial" part, which is really the most important part of the name. Mitochondrial DNA, which remains intact through the female line, makes it possible to identify ancestors all the way back through time. The same is true for the Y- chromosome, which remains intact through the male line.

To understand mitochondrial DNA and Y-chromosomes, start by thinking about how you'd create a genealogy of your ancestors. You'd get a piece of paper, write your name down, then draw a couple of lines from you to your parents, draw a couple of lines from each parent to their parents (your grandparents), and so on and so forth until you went as far back as you could go.

Now think about your DNA. You got half of your DNA from each of your parents, who each gave you one set out the two sets of chromosomes you have. So they each gave you half of their own DNA, but because chromosomes break and rejoin during the process of gamete formation, the DNA you got from each parent was a mixture of the DNA *they* got from each of *their* parents. Which is why one quarter of your DNA came from each grandparent, one eighth came from each great grandparent, and so on and so forth.

Now from the genealogical point of view for any given bit of DNA in your genome, it's not really possible to tell which of your great-great-great-grandparents that bit of your DNA came from — with two exceptions: your mitochondrial DNA and, if you're male, your Y-chromosome DNA as well. That's because these two kinds of DNA don't get scrambled every generation. You got your mitochondrial DNA from your mother; she got it from her mother, who got it from her mother, and so on all the way back down the line.

All of a sudden, you (or scientists who trace lineages through mitochondrial DNA) can go back in time from one mother to the next. Through mitochondrial DNA, a lineage can be traced through the maternal line.

The same thing is true of the Y-chromosome. Because in males it pairs up with the X-chromosome and they're different, it doesn't do any recombining. (X-chromosomes do recombine when they're in women, because women have two of them; men only have one). In males, the Y-chromosome never has a partner for recombination. If you're a man, you got your Y-chromosome from your father, who got it from his father, and so on and so forth. Scientists can trace the Y-chromosome back through the male lineage.

And that's why these are the bits of ancestral information we can get from our DNA.

Recombination doesn't happen with mitochondrial DNA or Y-chromosomes, but mutations can happen, and these accumulate through time — most are probably neutral because good mutations are rare and bad ones get weeded out — in different lineages. By studying these mutations, scientists can say which of us are more closely related. If you and I had a nearly identical mitochondrial genome, we would've had a most recent common great-great-great-great-etc. grandmother more recently than someone who's mitochondrial DNA was a little bit more different than ours.

Apply this bit of knowledge to all the people on earth. The more mitochondrial DNA diversity we have, the more we've all been diverging from a common mother (called Mitochondrial Eve) — and that's the information we use to figure out how long ago that woman existed. The deep branches of the tree suggests that our most common recent female ancestor lived in Africa a couple of hundred thousand years ago.

# Evolution within Homo Sapiens

Although the primary focus of this chapter is the evolutionary origin of the human species, human evolution didn't stop at the moment of speciation. Humans are still evolving. The following sections discuss some of the evolutionary events that have happened since the origin of our species.

## Natural selection: Still acting on humans

Because humans can alter the environment to suit ourselves, we sometimes assume that we've stopped evolving, but we haven't. Natural selection has continued to act, increasing the frequency of advantageous genes. In fact, the very changes we make can select for evolutionary changes. For some examples, read on.

## Of lice and men

If you always suspected that humans were special in some way, here's another bit of evidence to prove that you're absolutely right: Although most species of mammals have at most one kind of louse, the human species has three! This fact actually says quite a bit about human development.

Humans have head lice, adapted to hanging onto hair; body lice, adapted to hanging onto clothing; and pubic lice which like the nether regions. An analysis of the amount of genetic difference between the head lice and the body lice suggests that they diverged approximately 100,000 years ago, which gives scientists an approximate date of origin for tight-fitting fabric clothing — the sort body lice adapted to attach to. In the absence of clothing, lice were confined to hair, but when humans began to wear clothes, lice were able to spread to other regions of the body. Because the selective forces were different in each environment — head versus body — speciation between head and body lice eventually occurred.

So what about the third type of human lice: pubic lice? As it turns out, human pubic lice are closely related to gorilla body lice. They're divergent enough from gorilla lice, however, for scientists to be able to say that the transfer from gorillas to humans predated the evolution of body lice (and, therefore, the development of clothing). This finding suggests that humans lost their body hair (and, hence, had separate head and pubic-hair regions that different species of lice could colonize) long before they developed clothing.

Bottom line? Lice tell a story. If you have hair all over your body, you have only one kind of lice. Separate the hair into patches, and you can get lice adapted to each different patch. Cover some of the hairless area with a substance like clothing, and you can get diversification and speciation as one of the lice species (it just happened to be the one on the human head) radiates into this new habitat.

Oh, and by the way, lice are continuing to evolve. If you've ever had to deal with head lice and couldn't get rid of the buggers, you already know that many of them are now resistant to the chemicals we use to eradicate them.

### The evolution of lactose digestion

None of our primate relatives can digest *lactose* — the sugar found in milk — as adults. Infant mammals need to be able to digest milk, but historically, they haven't needed this ability after weaning. By domesticating dairy animals, however, humans altered the environment in a way that selected for evolution of lactose-digesting ability in adults.

Human lineages that are associated with dairy farming have a much lower level of lactose intolerance than do lineages that aren't. In Africa, for example, the Nigerian Yoruba, an ethnic group in Nigeria, are 99 percent lactose intolerant, whereas cattle herders in the southern Sudan are less than 20 percent lactose intolerant.

### Human disease and population density

The evolution of human disease is strongly affected by human population size. And large human populations became possible with advances in food production, namely the advent of farming. As human population density increased, human pests and parasites thrived, and humans evolved increased resistance to them. So in large populations, humans have developed built-in protection against the organisms that would do them harm.

But not all humans developed this resistance — just the ones living in high-density environments. Human populations living in low-density environments are relatively free of virulent pathogens. Small populations can't sustain infection, because everyone can be infected at the same time, and survivors of the disease become immune. The result? The pathogen goes extinct. If no one in the small group survives (in small groups, less chance exists that resistant individuals are present), the pathogen still goes extinct. As a result of the lack of constant exposure, small populations don't evolve resistance to disease. In a large population, on the other hand, new sensitive individuals are always being born.

When humans from large populations come into contact with humans who have traditionally lived in small groups, those from the large populations pass along diseases that can be far more serious for the people in the small populations. Tragic examples are the debilitating effects of disease on native populations in the Americas after contact with European explorers. The risk continues to this day when people from large populations come into contact with small groups in the Brazilian rain forest and other isolated locations.

Think about smallpox. Back in the day when smallpox was rampant, somebody always had it. (If at any instant no one was sick with smallpox, the virus would've gone extinct.) In a large population, new susceptible individuals were always being born, so the virus always had a refuge of new people to infect. Although smallpox was a dreaded disease, people in Europe, Asia, and

Africa were relatively resistant to it. Only 25 percent of people who contracted it died — a big number, true, but 75 percent of infected people survived. Compare that outcome with the experience of smaller populations, particularly in the New World.

The Americas were colonized via the Bering land bridge at a time preceding the development of agriculture and large population sizes. Movement to the New World would have been in small groups of hunter-gatherers — groups too small to maintain infectious diseases such as smallpox. As a result of the relative freedom from European diseases in the New World (an accident of history), human populations in the Americas didn't evolve resistance to these diseases and were devastated by them upon coming into contact with European explorers and colonists.

## Relaxation of selection

Allele frequencies can change in response to natural selection or in response to a reduction in selection pressures. As some of the selective forces that have acted on humans in the past are eliminated — humans no longer need to run away from big, fierce, hungry animals, for example — other alleles can increase in frequency as a result of genetic drift (random events; refer to Chapter 6).

In an environment in which running fast doesn't confer any particular benefit, *not* being able to run fast doesn't mean that humans are in any way less fit. As long as no tigers are around, our slow alleles don't have a lower probability of being passed on to the next generation. So we may get slower, but we don't get less fit. (Note that there isn't selection for slower running; there just isn't selection either way.)

## Cultural evolution

Cultural evolution isn't a biological phenomenon, but it's an interesting topic that can help you understand how humans are able to alter the environment rapidly and outcompete all the other species.

In biological evolution, genes get transmitted directly from parent to offspring, and they increase in the population as a result of either random forces or positive selection. Either way, the process can take some time. In cultural evolution, advancements (such as bright ideas or better ways of doing things) aren't limited to this vertical pathway of parent to offspring. If I figure out how to make a better spear, for example, I don't pass the information to my children genetically; I just show them how to do it. Then I show everybody else in the tribe how to do it, too.

Because cultural evolution can change behaviors or traits much faster than genetic evolution does, animals whose sole adaptive ability relies on genetic evolution are at a disadvantage. Rapid selection will occur to cause tigers to keep their distance from large groups of people armed with spears, for example, but changes in human hunting ability occur so suddenly that the tigers might still go extinct. Humans are still evolving biologically, but we're evolving culturally as well, which gives us an edge that other species can't match.

# Chapter 17

# The Evolution of Antibiotic Resistance

*H*ere's an interesting tidbit to throw out at your next dinner party: Our bodies have been colonized by all manner of bacteria. Yes, that's right; you are your own big blue marble. Fortunately, many of the bacteria that have set up housekeeping in your body are *commensal bacteria:* They colonize your skin or your gastrointestinal tract and are rarely of any concern. Others, however, such as *Staphylococcus aureus,* aren't such good citizens. They hang around on your skin and in your nasal (and other) passages, looking for ways to stir up trouble.

So why aren't you sick all the time? Because your immune system does a pretty good job of keeping potentially harmful bacteria at bay. And when your immune system has trouble, *antibiotics* — compounds designed to kill or neutralize infectious agents — can help. There's one hitch, though: Bacteria evolve in ways that make them better at overcoming our bodies' defenses and more resistant to the antibiotics we use to get rid of them.

In this chapter, I explain the evolution of antibiotic resistance in bacteria and show that the way humans use (and overuse) antibiotics is directly related to how bacteria evolve.

# Antibiotics and Antibiotic Resistance in a Nutshell

In this chapter, I use the term *antibiotic* to refer to any chemical that kills bacteria or inhibits bacterial growth. I'd stop there, but because the term *antibiotic* is often in the news and used interchangeably with terms like *antimicrobial* and *antiviral*, a little more explanation is in order — especially when these terms tend to appear in rather alarming news stories about infections running amok. To help you avoid confusion and gain a little perspective, I offer the following sections.

## Splitting microbial hairs: Defining antibiotics

Microbes are a diverse collection of organisms, many of which have almost nothing in common except that they are extremely little. They fall into three major categories — bacteria, viruses, and eukaryotes, which include fungi (which have some important disease organisms) — as well as a bunch of other critters, like Giardia and other protozoan parasites, that cause diseases like diarrhea and malaria. Humans tend to lump them all together because, as a group, they're responsible for infectious diseases.

*Antimicrobials,* as you may cleverly guess, are compounds that are active against microbes. Each antimicrobial has a name that indicates which microbe it's active against:

- **Antivirals:** Antimicrobial compounds that target viruses.
- **Antifungals:** Compounds active against fungi.
- **Antimalarials and other compounds:** These are active against specific kinds of eukaryotic diseases (see the nearby sidebar "Eukaryotes for you and me"). (***Note:*** Antimicrobials that are active against these types of infections tend to be named more specifically.)
- **Antibacterials:** Compounds that are active against bacteria.

Here's where matters get confusing: Antibacterials are often referred to as *antibiotics,* a term that is also occasionally used interchangeably with the term *antimicrobial.*

 *Antibiotic* can be used generally to refer to *all* disease-fighting compounds and specifically to mean *only* those compounds that are effective against bacterial agents. This dual-purpose use of the term leads to one of the biggest misunderstandings about what antibiotics can and can't do. You may have heard that antibiotics aren't effective against viral infections. That's true, when the term *antibiotics* is used specifically to mean *antibacterial,* which by definition means active only against bacteria, not against other microbes.

Technically, antibiotics are chemicals that kill or inhibit the growth of *biotic* (biological, or living) things, as opposed to abiotic things (like rocks). Rat poison is an antibiotic; so is Round Up. But when scientists use the term *antibiotic,* they're almost always referring to antibacterial agents — things that kill or inhibit the growth of bacteria — because the original antibiotic compounds, such as penicillin, were active against bacteria.

## A brief history of antibiotic resistance

In the days before antibiotics were widely available and widely used, people knew the dangers of infection. Even minor injuries like cuts and scrapes were taken far more seriously; just ask your grandparents. That's because an ounce of prevention is worth a pound of cure — all the more so when there isn't any cure.

Fast-forward to today. People tend not to view cuts and scrapes as being potentially serious medical conditions. Prevention seems less important, because we have a pound of cure — pounds and pounds and pounds, in fact. In 2007, people in the United States used millions of pounds of antibiotics. Therein lies the problem. With each passing year, antibiotics become less effective as bacterial populations evolve to be resistant to them.

Although recent news stories warning people about antibiotic-resistant strains of bacteria may lead you to believe that the phenomenon is relatively recent, it's actually as old as the use of antibiotics. Penicillin, the first widely used antibiotic, dates to the end of World War II. Within four years after its introduction, scientists observed resistant bacteria, and the incidence of resistance has increased steadily to the present day.

You'd think that having identified the fact that bacteria began to evolve almost immediately in response to penicillin would have encouraged people to be a bit more careful about the use of antibiotics. But we weren't, partly because it's hard to *not* use a medication that's so effective (many people considered penicillin to be a miracle drug) and partly because at the time, new antibiotic compounds were being discovered regularly. When one compound was no longer effective, doctors simply switched to a different compound. The scenario is very different today.

In recent years, bacteria have been gaining on us: The rate at which researchers have discovered new medically useful antibiotics has slowed, but the steady march of the evolution of resistance continues unchecked.

## Antibacterials, antifungals, antivirals, and you

Most of the antimicrobial compounds at humans' disposal are compounds that are active against bacteria rather than against other microorganisms, such as fungi and viruses. Why? Because developing compounds that are active against viruses and fungi is far harder. To be medically useful, a compound has to be able to stop the invading microorganisms without hurting the person taking the compound.

At the biochemical level, bacteria are quite different from people, making it possible to target specific details of the bacterial physiology without overly affecting human physiology. The same isn't true of viruses and fungi:

✔ Viruses replicate primarily by harnessing *human* cellular machinery. For that reason,

harming the virus without harming ourselves is very difficult.

✔ Fungi are quite a bit more similar to humans at the biochemical level than we might like to imagine; therefore, the substances that hurt fungi also hurt us (which is why it's so hard to cure a fungal infection like athlete's foot). Although chemicals are available that can harm the fungus pretty efficiently, the same chemicals tend to harm us, too.

This isn't to say that antibacterial compounds are without side effects. Many compounds do have side effects, which can be severe, but the medical risks of these side effects are balanced against the obvious benefits of curing the infection.

Again, the evolution of resistance isn't new. In every case, scientists have noted the existence of antibiotic-resistant bacteria shortly after the antibiotic was introduced.

Today, we humans now find ourselves facing bacteria that are resistant to many — and, in some cases, all — available antibiotics. Examples include staph, tuberculosis, syphilis, and gonorrhea. The most frightening thing we can observe from this information is that in the end, all of our antibacterial compounds end up being defeated.

# Becoming Resistant to Antibiotics: A How-to Guide

Try as scientists might to develop new and better antibiotics, bacteria are gaining on them, evolving resistance against every compound science can throw their way. Bacteria, in other words, are pretty good at this evolution thing, and here's why:

🖊 They reproduce extremely rapidly, and lots of them exist.

🖊 They've been at it a long time. Antibiotics are common in nature. Penicillin, the first antibiotic in wide use, comes from a fungus, *Pennicillium chrysogenum*.

🖊 They are the weapons bacteria and other organisms sometimes use in fighting each other.

Bacteria can gain antibiotic resistance in a couple of ways: by mutations and by gene transfers. The following sections give the details.

## Evolution via mutation

As I explain in Chapter 5, evolution by natural selection requires the existence of a variation on which selection can act. The initial source of all variation is the random mutations in an organism's DNA. These mutations occur during the processes of DNA replication (when copies are made) or repair. Basically, whenever an organism is doing something with its DNA, a chance exists that a mistake will occur, resulting in DNA with a slightly different sequence.

Because bacteria reproduce rapidly (sometimes as fast as every 15 minutes), and because so many of them are around, many opportunities are available for these changes in DNA sequence to occur. Put a billion bacteria in a beaker, and come back 15 minutes later; you could find 2 billion bacteria. In the course of duplicating those 1 billion genomes, a substantial number of errors will have occurred. Bacteria's fast, high-density lifestyle gives them an evolutionary edge.

You can observe this phenomenon in the laboratory very easily. An experimenter takes a single bacterium that's known to be sensitive to a particular antibiotic, drops the bacterium into a nice cozy beaker, and lets the bacterium divide. Pretty soon the beaker contains two bacteria, then four, and then eight. By the next day, the beaker contains millions and millions of bacteria, all descendents of the same antibiotic-sensitive parent. Now the experimenter takes these millions of bacteria and exposes them all to the antibiotic. Often, he finds that the beaker now contains resistant bacteria. Mutations have appeared in these bacteria as a result of DNA replication errors, and these mutations confer resistance to the antibiotic. When the remaining bacteria reproduce to fill the flask again, the experimenter ends up with a flask full of antibiotic-resistant bacteria.

Scientists would really like to develop an antibiotic to which bacteria can't evolve resistance, but so far, they haven't had any luck. The process of evolution is so powerful that when scientists change the bacteria's environment by adding antibiotics, they always manage to select for resistant bacteria. In our new antibiotic-drenched world, any bacterium that's a little bit better at surviving in the presence of antibiotics is going to be the one that leaves the most descendents.

A common misconception is that the addition of the antibiotic leads to the genetic changes that result in antibiotic resistance. But evolution requires that the variation *already be present.* The addition of the antibiotic didn't cause the bacteria in the beaker to become antibiotic resistant; it just killed all the antibiotic-sensitive bacteria, leaving behind only the bacteria that happened to be antibiotic resistant already.

## Evolution via gene transfer

An interesting characteristic of bacteria is that occasionally they acquire genes from bacteria of other species. Yes, that phenomenon is as weird as it sounds. Bacterial reproduction usually involves just dividing in two; no other bacteria is required, and both of the resulting bacteria are (excepting the occasional mutation) genetically identical. Every once in a while, however, a bacterium acquires genes from somewhere else. When it does, it's not too picky about which genes it gets.

An example of this kind of acquisition of new genes is the pathogenic *E. coli* 0157:H7 (see Chapter 15). *E. coli* 0157:H7 first came to light after an outbreak at a fast-food establishment; as of late, it's been found in a large number of domestic cattle operations, as well as in the occasional bag of spinach. This nasty version of *E. coli* started out as plain old, relatively harmless *E. coli,* but it picked up a whole bunch of genes that *E. coli* usually don't have, some of them quite nasty.

Gene transfer between bacteria can greatly speed the rate at which an antibiotic-resistant gene spreads. Rather than having to evolve separately in each individual bacterial species, antibiotic resistance can, after having evolved once, be transferred to different species.

This type of gene acquisition by bacteria (called *horizontal transfer* to differentiate it from *vertical transfer,* in which the trait is passed down through descendents), occurs via three mechanisms:

✔ **Transformation:** The process whereby a bacterial cell picks up DNA from its environment and incorporates it into its own genome.

✔ **Transduction:** The process whereby genes are carried from one bacterium to another in a bacterial virus. Occasionally, before leaving the host cell, a virus particle that's being assembled accidentally gets filled with bacterial DNA instead of viral DNA. When that virus particle latches onto a new bacterium, it injects that bacterium with the foreign bacterial DNA, which may then be incorporated into the bacterium's genome.

✔ **Conjugation:** The process in which two bacteria join and DNA is passed from one to the next. This process is controlled by a small circle of DNA living within the bacterium, called a *plasmid*.

Plasmids, which are much bigger than the pieces of DNA usually involved in transformation and transduction, are especially important in the spread of antibiotic resistance. Through a single plasmid, a bacterium can become resistant to numerous antibiotics in one fell swoop, sometimes with tragic results: An outbreak of Shigella dysentery containing a plasmid coding for resistance to four antibiotics was responsible for over 10,000 deaths in Guatemala during the late 1960s.

## Resistance at the cellular and biochemical levels

In the preceding sections, I talk about antibiotic resistance in general terms: how bacteria can evolve through mutations and through gene transfer. Whether resistance is conferred by mutations in DNA or by gene transfers, something goes on at the cellular and biochemical levels that changes the bacteria from being antibiotic sensitive to being antibiotic resistant. Basically, the resistance comes in one of three forms:

✔ Mutations that reduce the amount of antibiotic entering the bacterial cell, such as changes in the cell membrane that make it more difficult for the antibiotic to get in or that pump out the antibiotic as soon as it gets in.

✔ Mutations that enable the bacteria to produce enzymes that destroy the antibiotic.

✔ Mutations that change what the antibiotic targets. These changes can range from changing the shape of proteins so that the antibiotic no longer recognizes them to reorganizing biochemical pathways to eliminate the stages that the antibiotic targets.

## Mercy, mercy me! MRSA

Methicillin-resistant *Staphylococcus aureus* (MRSA) is a staph infection that can be very hard to treat because it's resistant to a lot of antibiotics including, no surprises here, methicillin, which otherwise would be wonderfully effective. According to the U.S. Centers for Disease Control and Prevention, MRSA infections (discovered in the early 1960s) accounted for 2 percent of staph infections in 1974. Thirty years later (by 2004), MRSA accounted for 63 percent.

MRSA is alarming for these reasons:

✓ Although MRSA staph infections can still be treated with other drugs, these drugs are more expensive, have greater side effects, and act more slowly than the drugs that were effective against it in the past. MRSA can often be treated with the antibiotic Vancomycin, for example. Unfortunately, we now have to worry about a new version of staph infection called, you guessed it, VRSA.

✓ MRSA infections cause thousands of deaths every year, but the most frightening ones, the ones that make for the most sensational news, are the cases of "flesh eating staph." Deep bacterial infections can cause massive tissue damage. While many kinds of bacteria have been implicated in such infections, MRSA is becoming an increasingly common cause. Because of the speed with which these deep infections can progress, treatment becomes a race against time. Even with the best medical care, fatality rates can be high, and having to use suboptimal antibiotics (because bacteria have grown resistant to the better ones) only makes the situation worse.

✓ Most MRSA infections (about 85 percent) are acquired in healthcare settings, but in the 1990s, MRSA began showing up in the broader community and are called CA-MRSA (for community associated MRSA).

# Evolving a bit at a time: Partial resistance

Bacteria may evolve complete immunity to the particular antibiotic by first passing through a stage of partial immunity. Partially resistant bacteria can survive a concentration of the antibiotic sufficient to kill susceptible bacteria but will succumb to a greater concentration.

Imagine a scenario in which no single mutation can render the bacteria completely immune to a particular antibiotic, but some mutations convey partial resistance. These mutations are favored only when antibiotic concentrations are low; such conditions lead to a population of partially resistant bacteria. Then another single mutation could result in the partially resistant bacteria becoming either more or completely resistant to the antibiotic.

Partial resistance is the reason why doctors tell you to finish a course of antibiotics — even if you're feeling better and even if you've decided that the last couple of pills are unnecessary. The initial pulse of antibiotics kills all the sensitive bacteria (which is why you feel better), but some of the remaining bacteria may be partially resistant to the antibiotic. Rather than let the partially resistant bacteria off the hook, don't give them a chance to get a leg up on the antibiotic. After all, they're the ones you really want to control. Stop them while you still can!

# The Battle against Antibiotic Resistance

Science is fighting a battle against antibiotic resistance — and losing because of how quickly microbes evolve. A big part of the problem is that we humans are the major selective force driving the evolution of antibiotic resistance in those critters that infect us. Every time scientists throw a new antibiotic at a microbe, they end up selecting for more resistant microbes. Science needs to change its strategy, and evolutionary biology can help:

- ✔ **Making new drugs:** If researchers understand exactly how the bacteria evolve resistance, they may be able to design better drugs.

- ✔ **Testing and refining theories about what happens if humans use antibiotics less:** The science of evolutionary biology, complete with all the beautiful experiments outlined in later sections of this chapter, is how scientists generate and modify their expectations. Unfortunately, much of what they learn from these experiments is bleak, but at least they learn something, and what they learn is important to understanding how microbial populations respond to changes in the way humans use antibiotics.

- ✔ **Not screwing up the drugs we already have:** By being careful about how and when we use antibiotic medications, humans can slow the development of antibiotic resistance.

## New and improved! Making new drugs

By understanding the mechanisms that make a particular microorganism antibiotic resistant, scientists can tailor new antibiotics to undermine that mechanism. If a bacterium is antibiotic resistant because it can pump antibiotics out of its cell, doctors can change the way the antibiotic is administered: Instead of giving the antibiotic alone, they can give it along with a second compound that disrupts each bacterium's pumping ability. *Voilà* — an effective solution. Similarly, by understanding exactly how bacterial enzymes destroy an antibiotic, scientists can design new, slightly different antibiotics that aren't as readily degraded.

# Turning back the clock to bacterial sensitivity

Some scientists have suggested that if humans temporarily cease using antibiotics, bacterial populations will lose their resistance in one of two ways — by evolving back to being susceptible to antibiotics or by being out competed by the remaining sensitive bacteria that are no longer at a disadvantage — at which point humans could resume using antibiotics with much greater effect.

## Evolving back

The thinking goes like this: Removing the antibiotic would result in the evolution of sensitive bacteria *only if* the sensitive bacteria have an advantage over the resistant bacteria. The most commonly envisioned advantage? That antibiotic resistance has a cost. Maybe the changes that make bacteria resistant (changes in bacterial membranes, for example, or DNA replication) also make them less able to perform other bacterial functions. By removing the antibiotic from the environment, mutations back to the original bacterial physiology would be favored, because sensitive bacteria are better than resistant bacteria at doing all the things bacteria have to do in the absence of antibiotics.

This hypothesis is great, but would it work? Several experiments have been conducted regarding this question, and they show, unfortunately, that the solution isn't as simple as instituting a moratorium on antibiotic use. Why? For the reasons explained in the following sections.

### Bacteria continue to evolve

Although bacteria resistance comes at a cost (they tend to grow more slowly in the presence of antibiotics than the original sensitive bacteria do), the process of *amelioration* — secondary mutations decreasing the debilitating effect of the initial mutation — is favored by selection. If the initial antibiotic-resistant mutant had a high fitness cost, subsequent evolution selects for additional mutations that increase the fitness of the antibiotic-resistant strain. As time goes by, the antibiotic-resistant bacteria don't have as low a fitness compared with the original bacteria (in the absence of antibiotics) as scientists would wish them to have. Thus, removing the antibiotics won't have as much of an effect as we would all want.

Starting with a wild sensitive bacterial strain, Stephanie Schrag and coworkers grew the strain until they could find and isolate an antibiotic-resistant bacterial mutant whose fitness was lower because of its mutation (the old DNA sequence was better than the new DNA sequence for some key bacterial function). They then grew this antibiotic-resistant bacteria for many generations in the lab and let evolution happen. As the antibiotic-resistant bacteria evolved, their fitness improved because of *compensatory mutations,* additional mutations that returned their fitness to about the level of the original antibiotic-sensitive strain.

Think back to that antibiotic-resistant bacterium that's not so good at DNA replication. Because it has the resistant allele, it manages to spread to the environment after all the sensitive bacteria have expired; now it lives in a sea of bacteria that are all resistant to the antibiotic but that all have a little trouble replicating their DNA. In this population, mutations that restore the ability to replicate DNA easily will be favored, and after selection has proceeded for some time, the antibiotic-resistant bacteria gain back much of the function they initially lost.

### Compensatory and back mutations aren't what they're cut out to be

So you have a strain of antibiotic-resistant bacteria that are now just as fit as the original antibiotic-sensitive strain. Now imagine what would happen if you take away all the antibiotics. Mutations still occur, and eventually one will appear that undoes what the original resistance mutation did (replaces the new sequence with the old sequence again), making this strain sensitive again. Will this new antibiotic sensitive strain take over so we can start using antibiotics again? Unfortunately no, because the old DNA sequence *is not* better than the new DNA sequence in the presence of the subsequent, post resistance, DNA changes. It was in the original strain, but not any more.

To determine that compensatory mutations could actually *lower* the probability that the resistant bacteria would evolve back toward sensitivity to antibiotics if antibiotics were removed, Schrag and coworkers conducted a second study. Using genetic engineering, they replaced the DNA responsible for antibiotic resistance with the DNA sequence of the sensitive strain while leaving the compensatory mutations unchanged. In effect, they created the bacteria that would exist if the antibiotic-resistant strain mutated back to being antibiotic sensitive

The surprising result was that the antibiotic-sensitive bacteria they'd created in the laboratory grew more slowly than the resistant strain in the absence of antibiotics. What this result means is that all the compensatory mutations that piled up in the antibiotic-resistant bacteria were advantageous only in the presence of the antibiotic-resistant gene; they were deleterious in the presence of the original sensitive gene.

After an antibiotic-resistant gene has appeared and compensatory mutations of the sort that Schrag and coworkers found in their laboratory have arisen, the antibiotic-resistant gene can be more fit than the antibiotic-sensitive gene, even in the absence of antibiotics. Why? Because a *back mutation* (a mutation that undoes exactly what the first mutation did) rendering one of these antibiotic-resistant bacteria susceptible could have lower fitness even in the absence of antibiotics. That situation is troubling because when all the bacteria become resistant, they stay resistant even if the antibiotic is stopped for a while. While we don't know exactly how common this sort of result would be, it's not good news.

### Mutations aren't necessarily as costly as you'd think

Although most antibiotic-resistant mutations have associated costs, some of them have very small costs, and some even have no cost. An awful lot of bacteria are around, and if even a few of them are capable of acquiring antibiotic-resistant mutations with no deleterious effects, this small class of mutations will be able to beat out the much larger number of antibiotic-resistant mutations that result in less healthy resistant bacteria. In a battle between the debilitated bacteria and the undamaged bacteria, the debilitated ones won't last long.

Just as genetic differences exist between different humans, genetic differences exist between different bacteria within a species. This situation raises the question of how important these different genetic backgrounds are in determining the associated costs of resistance. Fred Cohan and coworkers found that the genetic background in which an antibiotic mutation appeared was an important factor in the associated costs of that mutation. They determined that in some cases, acquiring resistance to an antibiotic was associated with few or no costs in other areas of bacterial physiology.

Cohan and his coworkers performed an experiment using the common soil bacteria *Bacillus subtilis*, which can take up DNA from the environment and incorporate it into its own genetic material. The researchers isolated a strain of *Bacillus subtilis* that was resistant to the antibiotic rifampin, made many copies of the gene that made the bacteria antibiotic resistant, and then introduced this gene into a collection of wild *Bacillus subtilis* strains simply by exposing samples of the different bacteria to the DNA responsible for antibiotic resistance. As they expected, the sample bacteria took up this gene.

For each of the wild bacteria, they compared the growth rate of the strain before the introduction of the antibiotic-resistant gene with the growth rate of the strain after the introduction of antibiotic resistance. In most, but not all, cases, they found a substantial decrease in growth rate with the introduction of the antibiotic-resistant gene.

Their conclusion? Antibiotic resistance indeed had a cost — but not always. Costs differed greatly among the different genetic backgrounds, and in a small number of cases, the antibiotic-resistant bacteria weren't significantly debilitated compared with the original sensitive strain.

### Antibiotic resistance and tuberculosis

To determine the clinical importance of antibiotic mutants with different fitness effects, Sebastien Gagneux, Clara Davis, Brendan Bohannan, and others conducted a series of experiments with the strain of bacteria that causes tuberculosis. They found that the mutations they could identify in the laboratory as having the lowest fitness costs were the same antibiotic-resistant mutations that were responsible for the resistant strains of tuberculosis present in clinical settings.

The researchers conducted laboratory experiments to measure the fitness costs of a series of antibiotic-resistant mutations. They examined a variety of mutations and found that some tended to have low fitness costs and that others tended to have high fitness costs. Then they examined the mutations present in clinical settings.

What they found was that antibiotic-resistant tuberculosis strains isolated from patients were very likely to contain antibiotic-resistance genes that had been shown in their laboratory experiments to have low fitness costs. They never found, in their patients, antibiotic-resistant tuberculosis strains with resistance alleles that had been shown to have high fitness costs. Both types of mutations occur, but the types with high fitness costs don't increase to the point where they are detectable. Furthermore, some of the clinical strains harboring the lowest-cost antibiotic-resistance mutation exhibited no decrease in fitness relative to the sensitive strain.

This is a finding of major importance, and unfortunately it's bad news. It had been suggested (and hoped) that if we reduced our use of antibiotics, the antibiotic-resistant bacteria would decline because they were supposed to be less fit than the original strains in the absence of antibiotics. They would just get overwhelmed by the few remaining resistant bacteria. Turns out that while many possible mutants are less fit, the ones that take over are the ones that aren't.

## Changing the way antibiotics are used

The results of the experiments outlined in the preceding section suggest that we humans had better think seriously about battling antibiotic resistance by trying to slow its progression rather than hoping we'll be able to deal with it after it occurs. To do that, we have to make real changes in how we use antibiotics — such as taking all our medicine. Taking only some of the antibiotics can create partially resistant bacteria, which may acquire subsequent mutations that make them even more resistant.

Here's just a little personal (and public) health note: Once you've knocked the infection down with the first half of the bottle, you've selected for the most resistant bacteria. The only way to kill those off is to finish the course of pills.

Another change is that people can use antibiotics less, which will decrease the selective pressure on the remaining sensitive bacteria. With fewer antibiotics around, sensitive bacteria would be less likely to be outcompeted by resistant bacteria. The idea is that, although humans can't stop using antibiotics to cure diseases, maybe we can stop hosing down the entire agricultural world with the same compounds we rely on to survive those diseases.

## Antibiotics and agriculture

Of the millions of pounds of antibiotics used annually in the United States, less than half are used to treat human diseases. The majority of antibiotics are used in agriculture and not primarily for treating sick animals:

✔ Low doses of antibiotics are used as a food additive to increase animal growth rate. The mechanism by which low doses of antibiotics increase animal growth rate is unclear, but what *is* clear is that farms and feedlots are growing reservoirs of antibiotic-resistance genes.

✔ Antibiotics are sprayed on fruit trees, resulting in a continuum of antibiotic concentrations starting from high at the center of the orchard and gradually fading to nothing at the edges of the antibiotic mist cloud. These continuous low doses of antibiotics in agricultural settings provide exactly the right situation for the evolution of fully resistant bacterial types.

Of the fraction of antibiotics in the United States that are used to treat human disease, evidence exists that many prescriptions are unnecessary. Patients want antibiotics even when they have no reason to think that their illness will respond to antibiotics — when they have the flu, for example — and the U.S. Centers for Disease Control and Prevention reports that more than 90 percent of physicians report feeling pressure to prescribe antibiotics that they know are unnecessary. So give your doctor a break!

For the past 60 years, antibiotics have allowed humans to live without fear of bacterial disease. With proper stewardship of this great resource, we can continue to enjoy the protections that antibiotics offer, but the point where we must take action to ensure the future efficacy of antibiotics is rapidly approaching. Scientists know with absolute certainty that bacteria will continue to evolve antibiotic resistance in response to human use of antibiotics. What is uncertain is whether we humans will make this evolution hard for them.

# Chapter 18

# HIV: The Origin and Evolution of a Virus

**S**mallpox. Influenza. Cholera. Syphilis. Yellow fever. Measles. Typhus. Plague. Malaria. Every age — and region — has its scourge, and some have more than their fair share. You may think that some these diseases are of the past (or of undeveloped countries) and that they don't really affect you. But when it comes to viral infections, no society or region is off the hook. Case in point: the appearance in the early 1980s of a new disease: *acquired immunodeficiency syndrome,* or *AIDS*.

AIDS scared the bejesus out of people. It was deadly. It was painful. And in the beginning at least, no one knew where it had come from or how it was spread. Fast-forward nearly 30 years. Scientists know quite a bit more about AIDS now than they did then. They know what causes it (the *human immunodeficiency virus,* or *HIV*), where it came from (other primates), and the path that the disease takes when it enters a human body.

How do they know all this stuff? By knowing how viruses work and by studying the evolution of the human immunodeficiency virus itself. Evolution is important in understanding HIV at many levels — from the origin of the virus to the disease process within a single patient to the development of effective treatment options.

# What Viruses Are

Given that this chapter and the one on influenza that follows both deal with viruses, knowing exactly what a virus is and how it functions is probably a good idea. To that end, viruses are

 ✔ **Microscopic entities:** The smallest microorganisms are viruses, even if not all viruses are smaller than all other organisms.

 Viruses used to be considered the smallest of all microorganisms (the smallest things are still viruses) until scientists found a really big one — so big, in fact, that you can see it under a light microscope. This Baby Huey of viruses is bigger than a lot of bacteria, and it infects amoebas.

 ✔ **Obligate intercellular parasites:** This is a fancy phrase that means viruses can survive and multiply only in living cells. Whose living cells? Why the hosts', of course.

A virus is an extremely simple organism. It consists of the viral genome (which can be made up of DNA *or* RNA) and any other molecules that the virus needs to get going after it's invaded the host cell. All this stuff is enclosed in an outer coat made of proteins or, in some cases, of proteins and *lipids,* which is a fancy word for *fat.*

## Are viruses alive? We say "Yes!"

Viruses are odd — so odd that no consensus exists on whether they're actually living things. Some people say yes; some people say no (and some people say sort of, but I'm not going to worry about the fence-sitters here).

One way to classify things as living or not living is to come up with a list of things that we think living things should have and then ask if our test object — in this case, viruses — has those things. Consider these examples of things that living organisms share:

 ✔ **Cells:** With the discovery of the microscope, which predated the discovery of viruses, scientists saw that all the living things they looked at had cells, so the definition of a living thing came to include having cells. Then scientists discovered viruses, which don't have cells. So under the "all living things have cells" definition, a virus, which doesn't have cells, can't be a living thing.

✔ **A genome made up of nucleic acid:** Viruses have a *genome,* the instruction set for making the organism, and when it's time to reproduce, they make new copies of the genome and package them into progeny viruses. The viral genome also codes for proteins, just as the human genome does, but because viruses are parasites, they don't need to make all the proteins — just the ones that the host doesn't already make. If having a nucleic-acid genome that codes for the production of proteins is the definition of life, viruses are alive.

If you say that, to be considered living things, viruses must have certain characteristics, then they are (or aren't) living organisms based on the characteristics you select. But this is a very nebulous way of defining what's alive and what isn't. If you say, for example, that having hair is a defining characteristic of a living organism, then you'd be alive, but a fish wouldn't. This example is silly, of course, but it shows how definitions have no foundations other than the ones humans give them.

A more helpful way to determine whether something is alive is to think about fundamental processes associated with life. Living things reproduce, for example. They also evolve. The process of copying the genome isn't error free, and mutations occur; mutations that are advantageous may increase in frequency. Being able to evolve is what fundamentally separates the living from the not living. A fire spreading through a forest reproduces itself, and a few flames lead to more flames, but fire itself doesn't evolve.

Viruses are independently evolving entities. They reproduce. They contain within themselves the instruction set for their own reproduction: the genome. And that genome changes through time; it evolves. If you use the life processes of reproduction and evolution as the fundamental characteristics of living organisms, viruses are alive.

The fact that viruses evolve is important not only for understanding where they come from and how they work, but also for addressing the threat that they pose and providing treatment. To treat a patient who's been poisoned, you just need to know how to counteract that particular poison. Treating a patient whose illness is the result of *an organism that can evolve* requires anticipating and responding to the adaptations that the organism can make. In the case of HIV, by the time a patient experiences symptoms, the virus has evolved and behaves differently from the original virus that infected the patient.

Whether you want to consider viruses living or not is entirely up to you. Regardless of which side of the debate you land on, all the information in this chapter still applies. Alive or not alive, viruses evolve, and *that's* the important thing.

## *Viral reproduction: DNA, RNA, or retro?*

The only way a virus can reproduce is to infect a host cell, hijack the cell's machinery to make copies of itself, and then move on to infect other cells. But not all viruses do things the same way, and these differences can have consequences for everything from viral evolution to treatment.

Different viruses use different nucleic acids (DNA or RNA) as the genetic material. Some, like herpes, use DNA. Some, like the flu, use RNA. And some (like HIV) use both, alternating back and forth. This third type of virus is called a *retrovirus.* Retroviruses use the enzyme reverse-transcriptase to make RNA from DNA (refer to Figure 18-1).

Different viruses have different mutation rates, and viruses that use RNA (either all or some of the time) can have very high mutation rates. HIV is one such virus. Its reverse transcriptase makes a lot of mistakes, leading to many slightly different viral types.

Some viruses replicate in the host cell and then the progeny go on to find another cell to infect. But others, including HIV, will also sometimes slip their DNA into the host chromosome. When the HIV genome is present as DNA, it can integrate into the host chromosome and hide there. It's just a string of A, C, T, and G bases and the immune system can't find it. Neither can current antiviral drugs. Hopefully, drugs that can find the virus can be developed. HIV researchers know what sequence to target; it's just a matter of figuring out how.

# *What Is HIV?*

*HIV* stands for *human immunodeficiency virus.* HIV attacks and impairs the cells in your immune system. HIV itself doesn't kill you; you die of any of the infections that your body usually would be able to fight off.

Several HIV viruses exist, and they can be divided into two major groups: HIV-1 and HIV-2. Both groups of viruses behave in a similar fashion, but the global HIV epidemic is primarily the result of HIV-1, and HIV-2 is confined mostly to one region of Africa.

The reason I say *groups* or *types* of viruses rather than *species* of virus is that scientists really have no idea what constitutes a viral species. Head to Chapter 8 for a discussion of what defines a species and why categorizing viral or bacterial organisms is so difficult.

Understanding the replication process that HIV uses is key to understanding why it's such a hard-to-treat disease. As the preceding section explains, HIV is a retrovirus; as such, it copies its RNA genome to DNA and then back to RNA. This replication process is important for two reasons, as explained in the following sections.

## Sneaking around in your chromosome

HIV does this really sneaky thing: When it copies its RNA genome into DNA, it integrates that DNA into the host chromosome (your chromosome). Now looking like just a few more nucleotides in the genome, it can hide there.

Your immune system can't find it as long as it's dormant and not doing anything. This phenomenon is also a big problem for HIV treatments that might be able to target the viruses — if it could find them.

## Attacking T cells

HIV targets cells of the immune system. Most important, it targets the *T cells* — white blood cells that fight infections either directly or indirectly. The type of T cell that is most susceptible to HIV infection is called the *helper T cell*. Helper T cells don't attack infection themselves, but they produce compounds that are involved in mediating the response of other T cells to the infection.

To do its dirty work, the virus looks around for something it can attach to, called a *receptor*. It finds the CD4 molecule on the T cell's surface and attaches to that.

Although the virus uses the CD4 molecule as a receptor, the molecule actually has some other purpose. Just as viruses hijack cells for their own purposes, HIV hijacks the CD4 molecule and uses it as an attachment point.

After attaching to the CD4 receptor, HIV attaches to a second receptor, called a *co-receptor*. HIV can use several co-receptors, but two are especially important in understanding the disease's progression:

✔ **CCR5:** This receptor is important in the early stages of infection. The form of HIV that attaches to the CCR5 receptor is called R5 HIV.

✔ **CXCR4:** This receptor is important later in the progression of the disease. The form of HIV that attaches to the CXCR4 receptor is called X4 HIV.

## Mutating like crazy

HIV is prone to mutations. Reverse transcriptase, which enables the virus to copy RNA back into DNA, is very error prone. This means that HIV has a high mutation rate, but because the virus is small, it can survive with such a high mutation rate.

The shorter the genome, the greater the probability that a sloppy copying process will be able to generate a copy without too many errors. (If you're a bad typist, you're more likely to correctly spell a short word than a long one, just because you have more chances to get something wrong in a long word.)

And even though most mutations are deleterious, so many HIV particles exist that an advantageous mutation (one that makes the virus resistant to a drug, for example) is likely to occur by chance.

# The History of the HIV Epidemic

Despite what some preachers said at the time, AIDS didn't drop out of the sky to plague the sinful or humble the proud. Instead, it was caused by a virus that simply did what viruses do. (Refer to the preceding section if you're unfamiliar with the general behavior of viruses.)

One thing that's so interesting about HIV viruses is that they're very similar to viruses that infect primates. Until this simian virus infected some unlucky person, AIDS didn't exist in the human population. Once in the human population, the previously simian virus was exposed to a new selective environment — namely us.

We don't know how often such cross-species transfers happen, but we do know (unfortunately) that sometimes the introduced virus heads down an evolutionary path that results in viruses that eat the new host for breakfast. In the case of AIDS, the primate virus can sometimes get a toehold in a human, and then selection sorts through all the various viral mutants, favoring those that are even better at infecting humans. Before such an event happened, AIDS wasn't a human epidemic; now it is.

Another interesting thing is that the primate-to-human transmission of the virus may have happened a whole lot earlier than people tend to think, given that most people think of AIDS and HIV as being "born" in the last decades of the 20th century. One of the reasons we noticed AIDS later may have been that people move around more nowadays. In the old days, you had to walk to the next village to spread whatever germs you might be carrying, and if people occasionally fell sick here and there due to an odd illness, not many people beyond their doctors or families knew about it. But today, with air-planes that can move people and viruses long distances in little time, a dis-ease can get all the way around the world in a day, and many more people can be exposed in a very short period of time.

# Where it came from

Many species of primates harbor viruses that are closely related to HIV. These viruses, called *SIV* for *simian immunodeficiency virus,* occur in 26 African pri-mate species. Both major groups of HIV — HIV-1 and HIV-2 — arose when the simian virus transferred from primates to humans. Scientists have been able to pinpoint which primate-to-human transfers resulted in both HIV-1 and HIV-2:

- ✔ **HIV-1 jumped to humans from chimpanzees, which harbor SIVcpz.** In at least three different events, SIVcpz jumped to humans, resulting in the three major groups of HIV-1. Each group represents an independent origin for the virus.

- ✔ **HIV-2 jumped to humans from sooty mangabeys, which harbor SIVsm.** At least half a dozen independent origins of HIV-2 from sooty mangabeys exist.

The naming conventions of the simian viruses indicate the species of primate that harbors the particular virus: SIVcpz stands for the chimpanzee simian virus, SIMsm for the sooty-mangabey simian virus, and so on. If your eyes tend to glaze over when you see what appears to be a random string of uppercase and lowercase letters, you may not have figured this convention out already.

## When a virus goes from one species to another

Sometimes, a virus in one species transfers to another species. In the case of HIV, a simian virus transferred to a human host, where it mutated into HIV. The phenomenon is fascinating and terrifying. People's fears about the bird flu — that after it's in a human host, it could mutate into a human version that's transmittable from human to human — have in essence already been realized with HIV viruses.

When people had no idea where HIV came from, they assumed a single origin. But when scientists started sequencing a bunch of simian immunodeficiency viruses, they realized that HIV (both 1 and 2) had leaped from primates to humans multiple times. Although scientists don't understand all that's involved in such a transfer, you shouldn't be too surprised. Given that humans are not all that different from primates (refer to Chapter 16), if we're going to catch a virus, getting it from other species of primates is as good a place as any.

### *The evidence of primate-to-human transfer*

How can scientists be so sure that the human immunodeficiency virus originated in simian populations? First, they know that humans can catch primate retroviruses. In at least one case, an animal handler acquired a simian immunodeficiency virus from a rhesus macaque (SIVmac), and cases of acquisition of other retroviruses have occurred.

Also, a strong correlation exists in Africa between the distribution of HIV types and the distribution of primates. The center of HIV-2 infection is the same part of Africa where sooty mangabeys live. And although HIV-1 has spread throughout Africa and the world, the origin of that virus appears to be the region inhabited by the subspecies of chimpanzees that harbors a related virus. These regions also provide ample opportunity for human-primate contact. Both chimps and sooty mangabeys are hunted for food, and young ones are occasionally kept as pets.

Other evidence includes

- **Gene sequencing:** The HIV and SIV viruses have been sequenced and compared, and the human viruses are remarkably similar to the simian virus that's considered to be its parent.

- **An HIV phylogenetic tree:** In reconstructing the tree (refer to Chapter 9), researchers discovered that instead of the HIV-1 strains appearing off to one side on their own little branch (which would indicate a single origin from one chimpanzee virus), the HIV strains are interspersed with the SIVcpz strains, indicating that HIV originated from three different simian strains.

  The same is true for HIV-2. Phylogenetic analysis doesn't show all the HIV-2 strains on their own branch of the viral tree, but instead several branches pop out of the sooty mangabeys viral tree.

## *A timeline*

HIV is a relatively new human disease. The U.S. Centers for Disease Control and Prevention (CDC) first reported on what we know now was the beginning of the AIDS epidemic in 1981. It wasn't until 1983 that the virus causing the disease, HIV, was identified.

Yet even though the disease was first recognized in 1981, it obviously was present before that time. Virologists decided to review old medical records and tissue samples to see whether anybody had suffered from the condition in the past.

People in the medical field often keep tissue samples in cases in which diagnosis proved to be difficult and an unknown disease seemed to be the culprit. Using modern molecular techniques in a process that can be thought of as a hunt for viral fossils, researchers were able to examine these historical samples for the presence of the human immunodeficiency virus.

You may be surprised to know just how early AIDS reared its head:

- ✔ The earliest known case of AIDS dates from a British sailor who developed AIDS-like symptoms in the late 1950s.

- ✔ The earliest known case of acquired immune deficiency in the United States (that is, the earliest date for which molecular evidence exists) occurred in 1968, involving a teenage male who reported that he'd been symptomatic for at least two years. Because he had never traveled outside the country, he must have caught the virus in the United States sometime before then.

- ✔ A survey of preserved blood samples has revealed antibodies to HIV dating back to the late 1950s.

- ✔ The degree of divergence between HIV and the suggested SIV source suggests that HIV mutated from SIV between 50 and 100 years ago.

So HIV was around possibly up to a century ago. A little more than 50 years ago, it presented itself in isolated cases. In the early 1980s, many people began to fall ill. What scientists don't know is whether the virus persisted at low levels until the later outbreak was noticed or whether it had gone extinct in the United States and then was reintroduced. What they do know is that by the end of the 1980s, AIDS was terrorizing the world.

# The Path and Evolution of HIV in the Patient

Every person infected with HIV has his or her own story, and the disease takes a dramatically different course in different people. Yet when you take all the sufferers in total, you get a picture — albeit a general one — of what HIV does when it enters the body.

Basically, right after infection, the levels of HIV in the blood increase rapidly, but then the body's immune system kicks in and is successful in reducing the amount of HIV virus present. Yet over time, the amount of HIV in the blood increases as the number of T cells decrease. Eventually, HIV destroys the immune system, eliminating T cells until almost none are left. At this point, the patient succumbs to infections that the immune system is no longer able to battle.

A key component of understanding the progression of HIV infection is understanding how the viral population evolves in the patient. The following sections explain these different evolutionary stages of HIV in the human body.

During the course of HIV infection the viral population evolves, first in response to the host immune system, then in the absence of immune response. Viral types unlike those in the initial infection arise during the course of the infection and are associated with later stages of disease progression.

# Increasing and growing more divergent

Right after infection, the levels of HIV in the blood increase very rapidly, yet the immune system is strong enough to fight back. As it suppresses the initial outbreak, the infected person's immune system also exerts a powerful selective force on the HIV virus.

The infected person's immune system recognizes HIV as a foreign invader and attacks it. But HIV has a high mutation rate, and not all of the individual viruses produced in the infection are exactly same, so some are able to avoid being targeted. So now the immune system has to go after these slightly different viruses and, in doing so, selects for viruses that are even more different. This cycle—in which the attacks launched by the immune system attacks select for slightly different viruses—goes on, back and forth for quite some time. The immune system puts up a valiant fight, often for years. During this period the HIV population evolves in two ways.

### The viral population becomes progressively different from the original strain

The viral population evolves to be progressively more different from the original infecting strain. Initially, the person's immune system is pretty good at battling the virus. So although HIV increases immediately after infection, it drops back down when the immune system fights back.

But as the disease progresses, the HIV population in the person evolves. Mutant virus progeny that are less susceptible to the immune system increase in frequency. Over time, the virus population becomes more and more different from the original infecting particles. The viral population is evolving in response to the immune system — the immune system is fighting hard and only new viral mutants can avoid it.

### The virus population grows increasingly diverse

The virus population as a whole becomes increasingly diverse — that is, there are progressively more different HIV types within the patient as the disease advances. The most fit HIV viruses will be the ones that are different from those that the immune system has responded to, but there are many ways to be different so viral diversity goes up.

# Reaching a plateau

HIV destroys one of the key components of the immune system, a type of T cell. When enough of these cells have been destroyed, the immune system can no longer contain the virus, and the HIV population starts to evolve in a new way.

As T cells are destroyed, the crippled immune system no longer acts as an agent of selection on the HIV viruses. At this point, selection no longer favors viruses simply because they are different — they don't have to escape the immune system, which is worn out and can fight no more. The difference from the original strain reaches a plateau. Mutations still occur during HIV replication, but now if a particular strain is especially good at growing in the host, there's no immune system to knock it back down.

# Destroying T-cells in a different way

HIV disease progression can vary dramatically from patient to patient, and evolution does not occur along exactly the same trajectory in each case. But one major event which is observed repeatedly is the evolution of viruses later in infection that attack the immune system in a different way. These viral variants (called X4 variants) are able to attach to a different part of the surface of immune cells (a cellular feature called the CXCR4 receptor ) and destroy the T cells in a different way.

The X4 type begins to increase in frequency around the time that the overall genetic diversity of the viral population is decreasing, and the peak of the X4 virus type is associated with period proceeding transition to full blown AIDS symptoms.

Researchers don't completely understand how theses X4 viruses influence the disease or whether the X4 variants evolve in every patient. What they do know is that the X4 variants cause T cell death differently from the original variants.

# Other Interesting Facts about HIV

HIV is as interesting as it is scary. Consider this to be the section of other cool things about the human immunodeficiency virus — if you can use the word *cool* when talking about a deadly pathogen. If you can't, consider this to be the section of noteworthy-but-not-vitally-important information for serious-minded readers.

## Some people may be resistant

During the initial infection, the virus binds to a CCR5 co-receptor (refer to "Attacking T cells," earlier in this chapter). As it turns out, not everyone has that same molecule. Some people have a mutation that makes a slightly different CCR5 molecule. People who have one mutated copy of the CCR5 locus show delayed progression toward full-blown AIDS. And people who have two mutated copies appear to be resistant to HIV infection.

Interestingly, this mutation is found primarily in Europe, and its prevalence is correlated with the degree to which a particular location experienced plagues in the past. It's been hypothesized that this mutation may have made people more resistant to certain other disease organisms and that as a result the mutation increased in frequency in places that once experienced plague. Just by coincidence, the same mutation confers resistance to HIV infection.

Scientists aren't sure that the bubonic plague was caused by black-plague bacteria. Instead, it may have been caused by a viral hemorrhagic fever.

## HIV evolves in a new host

Evolution within the patient selects for a more divergent virus, and that viral diversity increases through time in the individual (refer to the earlier section "HIV Evolution within the Patient"). When you compare the virus present during the initial period of infection across different people, however, you find less divergence among people than you see over time within one person.

Think about this for a minute: If the virus is diverging in one person, and that person infects a second person, you'd expect that the second person would start out with a more divergent population of the initially infecting virus. But that's not the case.

Here's why: After infection, strong selection for the ancestral viral type occurs, so the virus evolves back toward this condition. This selection may happen for at least a couple of reasons:

✔ The viral forms present in a person late in the infection may not be well suited to living in the environment present in a newly infected person. Hence, selection favors going back to an initial viral form.

✔ Although the diverse viral population may be perfectly capable of surviving in a newly infected host, this population may be outcompeted by viral variants that have mutations resurrecting the ancestral type.

## HIV has a high recombination rate

One way that the HIV population generates all that diversity is via a huge amount of recombination among viruses. In *recombination,* a DNA or RNA sequence (whichever the critter uses for genomic information) is produced that's a combination of two original sequences. For most organisms, *recombination* is generally believed to happen far less frequently than mutation. But in HIV, the recombination rate may approach the rate of mutation.

Scientists know about the recombination rate by making and analyzing phylogenetic trees for a group of viruses with different genes. For organisms with low rates of recombination, phylogenies constructed from different genes should have trees that match. The extent to which they don't match allows scientists to determine the frequency of recombination.

When you take a bunch of HIV viruses, determine the sequence of several genes, and then make phylogenetic trees from these different genes, the resulting trees don't always match. This shows that, rather than being a rare event, recombination is a major source of the diversity in the HIV population within a patient.

# Using Evolution to Fight HIV

The major problem we face in coming up with a vaccine for HIV is that the virus mutates so rapidly that it ends up being different enough from the strain used to make the vaccine that the vaccine doesn't confer immunity. Scientists can use the techniques of phylogenetic analysis to study the evolution of HIV and identity which parts of the HIV virus are more or less likely to change, and it might, and I stress *might,* be possible to use this information in our hunt for an effective vaccine.

Morgane Rolland, David Nickle, and James Mullins at the University of Washington have been working on just this problem: how to design a vaccine that essentially sneaks up on the virus when it's not looking. They hypothesize that the trick might be to eliminate all the info from the parts of the HIV genome that are mutating most rapidly and focus on the parts that aren't. The key is to figure out what parts of the HIV virus have been outwitting the immune system and then design a vaccine that targets parts of the virus that can't evolve.

By understanding how the virus evolves, we may be able to get a few steps ahead of it and make a vaccine that's harder for the virus to evade. This would be a remarkably cool application of evolutionary biology to a real-world problem.

# Chapter 19

# Influenza: One Flu, Two Flu, Your Flu, Bird Flu

## In This Chapter

▶ Discovering where new flu strains come from

▶ Predicting next year's flu today

▶ Understanding vaccines

*I*nfluenza. Before the mid-20th century, people gave this disease the respect it deserved. Now we call it the flu; act fairly cavalier about the symptoms unless we need a break from work; and poo-poo the vaccine unless we're in a high risk group populated by babies, octogenarians, and hypochondriacs. Periodically, though, we get reminded that the flu can be serious business — a fact that epidemiologists and virologists have been trying to beat into our thick skulls forever — and then we fly into a panic.

Well, here's news for you, some good, some bad:

✔ Many strains of influenza exist, and courtesy of evolution, even more are on the way. Their rapid rate of evolution makes them a constant problem, and every so often, an especially nasty strain catches scientists by surprise.

✔ Although some of us may succumb, we aren't completely helpless. Scientists have been fighting back, even using the process of evolution to turn the tables on these pesky viruses. Thanks to techniques for vaccine design, scientists are getting closer to being prepared for these new and improved strains as well.

This chapter has the details on all the bad things the flu can do and all the reasons by you don't need to panic — at least not yet.

# The Flu and Your Immune System

Most likely, your immune system has already been introduced to the flu. You got sick, felt lousy for a while, and eventually got better. Why? Because your body has an immune system. To understand the rather eventful history of influenza in the human population, you need to know how the flu virus infects humans and how the human immune system fights back. Consider it a play in three acts.

## Act 1: The virus attacks and spreads

The influenza virus, which causes the illness commonly referred to as the flu, is most often transmitted from one person to the next in small droplets drifting through the air — possibly the result of somebody's sneeze or cough.

One day, you're unlucky enough to inhale one of these droplets, and a flu-virus particle attaches to one of your cells. Once inside the cell, the particle takes over the machinery necessary to make more copies of itself. Then your cell bursts and releases all these copies, which go looking for more cells to infect. At this point, you're sick, and you'll probably help the virus get to new cells in other people when you start coughing. End Act 1.

## Act 2: The body fights back

You're sick, but you don't stay sick. Instead, your body responds to the invading flu, fights back, and overcomes the infection. How? Keep reading.

Your body has numerous, quite complicated systems for fighting off microscopic foreign invaders such as the flu virus. Combined, these various mechanisms make up your *immune system*. I won't get into the details here except to note that your body has two general classes of response:

- **Nonspecific responses:** Nonspecific responses are reactions, such as tissue inflammation and fever, that are generally bad for all invading microbes. When your body senses the presence of invading microbes, it sometimes turns up the heat in its fight against them. Indeed, having a fever may make you feel bad, but it makes the invading microbes feel even worse.

- **Specific responses:** *Specific responses* include such things as *antibodies,* which are special proteins that seek out and destroy invaders. When your body becomes aware of the flu, the first thing it notices is that the flu virus is foreign. It's not part of you and, as far as your body is concerned, doesn't belong in you. That's why as soon as the flu has been identified as "not self," your body begins to produce antibodies that specifically target this strain of flu.

As your body responds to the invader in both nonspecific and specific ways, it starts winning the battle against the flu. After a week of two, all the flu particles in your body have been killed, and you're on the road to recovery.

## Act 3: Building up the guard

Your immune system has a very special property called *memory*. Although it takes a little time for your body to produce enough antibodies to beat back this strain of flu, memory means that after your body has responded once, it's immediately ready to respond again.

That's great news for you, because the flu strain you just recovered from can no longer invade your body; now you are immune to it. And since your body's antibodies are able to attach to different but similar flu strains, you also have resistance to those as well.

### Which leads to a sequel: The return of the flu

The evolution of influenza is strongly affected by the immune system's response. The strain of flu from which you recovered was clearly quite good at being the flu. It managed to get not only from some other person into you, but quite possibly into a third person as well. In fact, it's possible that the strain of the flu you got also spread through your workplace, your kid's school, your town, the rest of the country, and even the world. Indeed, flu strains do this all the time. But after everyone who had a particular strain of flu gets better, all of them are immune to that strain. Thus, the next time you get the flu, it will be a genetically different flu strain.

## The Three Types of Influenza: A, B, C

Three different strains of flu infect humans. These strains are conveniently named influenza A, influenza B, and influenza C. They all have genomes that use RNA instead of DNA as the genetic material:

 ✔ **Influenza A:** This flu sweeps through the human population each flu season. It evolves so fast that strains are sufficiently different to avoid existing antibodies every year. Influenza A occurs in humans and a variety of other animals, which makes the evolution of influenza A somewhat more complicated, as well as more interesting for this discussion. The remainder of this chapter concentrates mostly on influenza A.

✔ **Influenza B:** Influenza B is just like influenza A, except that influenza B exists only in humans. Beyond this single difference, you just need to remember that influenza B evolves similarly to influenza A.

✔ **Influenza C:** Because most people are immune to influenza C, this strain of flu is not characterized by epidemics that periodically sweep through the human population. Instead, it is primarily a disease of the young, because it has a chance to reproduce only in children who have not yet developed an immunity to this generally mild flu. I mention it here in case you're curious, but it really isn't important to the discussion in this chapter.

Unlike influenza A and B, influenza C evolves very slowly. We don't understand exactly why it evolves so slowly, but as result of this slower evolution, all influenza C strains are basically the same. About 97 percent of Americans have antibodies to influenza C, which means it's likely that at some point in your life, you were infected with it and developed a resistance to it. Now you are immune to the strain that infected you, and as a result of the high level of similarity among strains, you are also immune to all other influenza C viruses.

# The Evolution of Influenza A

Influenza A is a very small virus with a genome made of RNA instead of DNA. Its genome is comprised of eight different RNA segments, and it has only ten genes. (For comparison, consider that people have over 25,000 genes.) Influenza A is a remarkably compact organism, and these ten genes enable it to do everything it needs to do to infect you and reproduce.

## Mechanisms of evolution: Mutation, recombination, or reassortment

The influenza A virus (and the influenza B and C viruses too, for that matter) can evolve in three ways:

✔ **Mutations:** Influenza is a virus with an RNA genome, and RNA replication has a high error rate. For this reason, the virus mutates rapidly. That's why this year's flu is different from last year's flu. The strains may be similar enough for the immunity you developed to last year's flu to protect you (or at least partially protect you) but don't count on it.

✔ **Reassortment:** Because influenza has a genome divided into segments, new variants can also arise via *genetic reassortment,* the process whereby two different viruses infect the same cell, and the progeny have segments that are a combination of these two strains. (For an explanation of how viruses can reproduce this way, refer to Chapter 12.)

When reassortment involves the segments that code for the H or N proteins that cover the outside of the flu virus, the new strains (or the strains produced) can be so different from the parent strains that no host has even partial immunity. As a result, these strains can sweep rapidly through entire populations.

✔ **Recombination:** In *recombination,* one of the existing influenza genes is incorrectly replicated and ends up with a new section of RNA spliced into it. This sequence could come from another influenza strain or even from a host cell's RNA. Although such events are almost always likely to be harmful to the virus, every once in a while a new sequence with foreign RNA spliced into just the right part of one of the surface proteins may make the new flu strain resistant to existing host antibodies.

## Genes to know

For the purposes of this discussion, you need to pay attention to only four genes: three on the outside and one that determines which species a particular strain can infect.

The three genes on the outside are

✔ Hemagglutinin (abbreviated H)

✔ Neuraminidase (abbreviated N)

✔ Matrix

The fourth is the Nucleoprotein gene (also called *Nucleocapsid*) and is abbreviated NP.

Hemagglutinin, neuraminidase, and matrix code for proteins that are on the surface of the influenza particle. These surface proteins are important because when the flu enters your body, these proteins are the ones that your immune system can potentially see — and guard against.

The Nucleoprotein gene is important is determining the *host-specificity* of a particular flu strain — a fancy way of saying which critters that particular strain can infect.

### The importance of H and N

As noted in the preceding section, hemagglutinin and neuraminidase are abbreviated H and N, respectively. If you've been reading about the flu or hearing about it on TV, you may have heard a particular strain described in terms of H and N and some numbers — for example, avian influenza H5N1.

There are 16 different hemagglutinin proteins and 9 different neuraminidase proteins, but each individual influenza A virus has only 1 type of each protein. The numbers in a strain's name (H**5**N**1**, for example) indicate which neurominidase and hemagglutinin proteins that particular strain has.

All the H and N types occur in waterfowl, but only some of them occur in humans. At this time, two types of influenza A are circulating in the human population: H3N2 and H1N1. Ideally, avian strain H5N1 — the *bird flu* you occasionally hear about in the news — won't mutate to easily infect humans and be transmittable between them, and make it three. Right now we can catch this strain from handling infected birds, but we don't seem to pass it to other humans, so it can't spread in the human population. At least not yet.

### NP: The host-specificity gene

From the earlier section as well as the reports in the newspapers these days, you know that humans aren't the only ones plagued by the flu. Birds get the flu, too, specifically waterfowl like ducks and geese. But people and birds don't usually get the same strain of the flu because flu strains tend to be species-specific. A human cell is different from a goose cell, and the flu strains that replicate in geese don't usually infect humans. In short, flu strains tend to be specific to a particular groups of animals. The gene that controls (in large part) which species of animal a flu strain can best infect is the NP gene.

The H and N types determine how readily your immune system can see the flu. If a particular strain is an H-N type that your body has seen before, then you've got a jump on it. If it's one your body hasn't seen before, then the flu has the jump on you. The NP gene determines whether the flu can see you, that is, whether it can reproduce in your body. The worst possible combination is an NP gene that lets the flu chow down on you wrapped in surface proteins that take your immune system by surprise.

## Who gets the flu and from where

There are five major groups of Influenza A strains. In addition to the strains that infect humans and waterfowl (ducks and geese), there is a group of flu strains that infects horses, one that infects pigs, and one that infects seagulls (which, contrary to what you may reasonably assume, are *not* waterfowl).

By taking a closer look at the genetic sequences of the NP gene (the one that determines which species a flu strain can best infect), scientists find that it's usually easy to tell the human flu strains from the waterfowl flu strains from horse flu strains and so on because flu strains are species specific. Seagulls tend to get seagull flu, horses tend to get horse flu, pigs tend to get pig flu, and so on. But not always.

One of the most interesting conclusions that can be drawn from an analysis of the sequences of Influenza A is that, much as we tend to think of this little beast as a human virus, its home base appears to be waterfowl. It seems that every so often, a flu strain jumps from waterfowl to some other susceptible species where it persists for a while and then goes extinct and/or is replaced by another strain. All the seagull, human, pig, and horse influenza A strains currently in circulation were probably acquired from waterfowl within the last 100 years. Understanding what causes these events is particularly important today, when public health officials and governmental agencies are coping with the very real possibility of another devastating flu pandemic.

### The evolutionary history and relatedness of different flu strains

The RNA sequence information allows scientists to reconstruct the family tree of the flu and figure out how the five major groups of the flu are related to each other. Humans, for example, have influenza strains similar to those in pigs; in fact, it's not unheard of for pigs to catch the flu from people, and vice versa. (The Swine flu scare of 1976 is one example).

In addition to helping scientists formulate hypotheses about the relationship between the different flu groups, surveying the RNA sequence variation for a collection of flu strains allows them to measure the amount of variation in the different groups of the flu. This information can provide information about underlying evolutionary processes.

An important finding is that not all groups have the same amount of variation in the host-specificity protein. A collection of human flu strains has quite a lot of variation, as do the flu strains that attack horses, pigs, and seagulls. But the flu strains infecting waterfowl are much more uniform at the protein level. The waterfowl strains have the greatest diversity of surface proteins — all the different H and N molecule types — but when it comes to the proteins important for actually chowing down on ducks and geese, most of the different waterfowl flu strains are about the same. There just isn't very much variation at the protein level, even though there's variation at the RNA level. (Remember that there could be changes at the nucleotide level that don't affect what protein is made.)

Scientists know that mutations are common in RNA replication, but when they look at the host specificity gene of the flu strain that infects waterfowl, they find that most of the variation is neutral; it doesn't affect proteins. Mutations that affect protein structure are certainly always appearing, but in the waterfowl population, it seems these mutations don't increase in frequency. We know they happen, but we don't see them. From this, we can speculate that the flu strains that are infecting waterfowl have, at least for now, gotten to be about as good as they can be.

## Duck, duck, goose: How waterfowl flu spreads

Interestingly, the waterfowl flu strain doesn't usually tend to make waterfowl very sick, and it doesn't have to be spread from one duck or goose to the next. Waterfowl are really good at making bird droppings (if geese or ducks have ever taken up residence at your local park, you know what I'm talking about), and these droppings contain huge amounts of flu virus. The flu in ducks lives in the gastrointestinal tract, not the lungs, and those droppings spread the flu virus through the entire pond where it can easily infect many other ducks.

Every once in awhile, some other organism comes into contact with some of the waterfowl flu — easy to see how that can happen when you think of a pond full of influenza infected droppings. When this happens, the chance exists for the waterfowl flu to infect a different kind of animal. Scientists have evidence of several different cross-species transfers from ducks to a variety of mammals.

### Which strains are evolving faster

Because of differences between neutral and non-neutral sequence variation, scientists can tell that the flu strains in waterfowl are evolving differently than the flu strains in the other groups. Although all five groups have variation at the nucleotide level, the waterfowl group doesn't have any significant variation at the protein level. The flu in waterfowl seems to be at an evolutionarily stable point, evidence that selection is acting differently in different groups of flu.

Take-home message: The host-specificity protein in waterfowl has evolved to some optimum level. Think of it as a peak on the adaptive landscape. (If you don't know what the adaptive landscape is, refer to Chapter 6.) Mutations that occur that change the protein are just as rapidly eliminated from the population, so we don't see them. What we do see are changes in the RNA that don't affect the protein.

### Jumping from one species to another

Because different species tend to be infected with flu viruses specific to their species, scientists who collect flu strains simply have to know which organism it came from to make a pretty good prediction about which strain it is. Take a strain from a horse, for example, and analyze its genetic sequence, and chances are, it'll be recognizable as a horse flu. Sometimes, however, scientists find a sick animal, collect the specimen, do the analysis, and find an unexpected flu strain.

When an animal has been infected with a flu strain different from the one expected, scientists use phylogenetic analysis again, because in making the tree of the five groups they can spot when a particular strain is out of place. If they find that a horse is sick with the flu, for example, they only have to check the flu sequence to discover whether it's the strain they'd expect (one specific to horses) or one they didn't expect (a strain that horses don't typically get). They can also tell in what ways it's different: that it's way different from the horse flu, for example, and just like the bird flu.

Such sequence analysis tells scientists that sometimes flu from waterfowl can infect other groups. Its been shown to move to horses and pigs, as well as other animals not usually associated with the flu, such as whales, seals, and mink. Now of course we're finding these flu strains in unlucky people in Asia as well.

That waterfowl flu can be transferred to other species is significant because the waterfowl population contains the highest diversity of surface proteins. Were a change to occur in a waterfowl flu that would allow it to reproduce in human cells, for example, it could take our immune system by surprise. The strain wouldn't look like any of the other flu strains that your immune system has responded to before. Although your immune system would respond to the infection, it'd be starting from scratch. If the flu strain is particularly virulent, you might not have that much time.

Bottom line. The waterfowl flu population is a source of cross species infection, and it's implicated as a source of new variants that have infected the human population in the past, a process we may be seeing the early stages of with the H5N1 avian flu.

# Learning from the Past: Flu Pandemics

The term *pandemic* refers to an epidemic of an infectious disease that occurs over a large area, such as a continent or even the entire world. The black plague that swept from Asia through Europe in the mid-1300s is a good example of a pandemic. Flu pandemics can occur when a flu strain that appears in the human population is so different from the circulating strains that the worldwide population is completely susceptible. As a result, this new flu is able to sweep rapidly through the entire world. Often, influenza pandemics are given names related to the regions where the virus first appeared.

In an average year, the Centers for Disease Control and Prevention estimates that influenza kills more than 20,000 Americans. The global total is harder to estimate accurately, but the World Health Organization places it at somewhere between 250,000 and 500,000 deaths per year. Pandemic years see a much, *much* higher death rate — sometimes into the millions.

Scientists have information on flu pandemics dating from the pandemic of 1889 to the more recent 1977 pandemic. By studying the genome sequences of the more recent strains and the antibodies of the victims in the more distant strains, researchers have been able to piece together the origin of these pandemic strains. The following sections take you on a little stroll through the pandemic flus of the last hundred years.

# Pandemics of 1889 and 1900

Because sequence information isn't available, scientists have to rely on *seroarchaeology* — the search for antibodies in preserved tissue samples — for these pandemics. The antibodies in these tissue samples can't tell researchers about all the genes in the flu strains responsible for the pandemics, but they *can* identify the H and N proteins, because these proteins are the external parts of the flu virus that the body's immune system produces antibodies against. So we know that the 1889 and 1900 pandemics were caused by an H2N2 and an H3N8 strain, respectively.

# The Spanish flu (1918)

This influenza pandemic strain is the first one for which researchers have complete sequence information. Sequence analysis of the genes from this strain suggests that it was the result of an H1N1 avian influenza virus entering the human population. This finding is important, particularly in light of recent concerns about the possibility that the rare but often fatal H5N1 avian flu will change to a form that can be easily spread between humans. The Spanish flu is an example of a flu strain from one host (waterfowl) infecting a new host (us).

# The Asian flu (1957)

The Asian flu strain (H2N2) contained 5 gene segments from the circulating 1918 strain and 3 new ones — the different versions of the H and N genes, as well as a gene involved in RNA replication. These changes were almost certainly the result of a reassortment event with an avian strain of influenza A, and the new strain replaced the H1N1 strain from which it was derived.

The Asian flu pandemic was the result of a reassortment event between the original H1N1 strain and an avian strain that contributed the 3 new genes. These two strains ended up in the same host, the various segments got mixed up during the flu's replication, and out popped something new.

# The last great pandemic

The 1918 Spanish-flu pandemic killed more people than any other disease outbreak in human history. It is estimated that between 20 million and 40 million people died of the flu during the 1918–1919 flu season. Furthermore, although influenza is usually most fatal among the very young and the very old, the 1918 flu had very high mortality across all ages.

It was not until very recently that scientists learned the complete genome sequence of the 1918 flu. No samples suitable for genome sequencing were known before that. However, in 2005, viral nucleic acids were isolated from the bodies of victims who died in Alaska during the epidemic. Their bodies had been buried in the Alaskan permafrost and had remained frozen until the present. These frozen samples were sufficiently preserved that scientists could isolate the viral genetic material (RNA) from them.

Before the rediscovery of the 1918 flu, scientists thought that the 1918 pandemic originated in a fashion similar to the 1957 and 1968 pandemics — that is, via the acquisition of several new genome segments from a reassortment event with an avian strain. Scientists now know, however, that all eight segments of the 1918 strain are very similar to avian strains, and the current hypothesis is that the 1918 pandemic was the result of an avian strain entering the human population.

# The Hong Kong flu (1968)

When the Asian flu strain (H2N2; see the preceding section) went through a second reassortment with a different avian influenza strain, the H3N2 strain was born. This strain had the same N gene but had acquired 2 new genes, including a new H gene.

This is another example of a reassortment event between a human strain and an avian flu strain. This H3N2 strain replaced the H2N2 strain in the human population.

# The Russian flu (1977)

In 1977, the H1N1 flu last seen in humans in 1950 reappeared. Because most people older than about 20 had been exposed to H1N1 flu, the resulting epidemic was not severe enough to be labeled a true pandemic, but it was certainly a significant event.

Scientists don't know exactly where this strain came from, but it has been suggested that it escaped from a laboratory somewhere. Support for this hypothesis is not just the strain's sudden reappearance, surprising though that is. More convincing is that the 1977 H1N1 was almost identical to the 1950 H1N1 — in all those years, this strain hadn't been evolving.

Remember that this year's flu isn't quite the same as last year's flu, and next year's flu won't be the same as the flu that follows it. The point? Twenty-seven years without change is hard to explain — hence the suggestion that maybe those 27 years were spent in a test tube in a freezer!

Since this event in 1977, two influenza A lineages have been circulating simultaneously in the human population: H3N2 and H1N1.

# Fighting Back: The Art and Science of Making Flu Vaccines

Vaccines work because of immune-system memory, which essentially prepares your immune system to defend you immediately when it senses that a previous invader — or one like it — has returned for a repeat performance. (Refer to "Act 3: Building up the guard," earlier in this chapter, for details.) By combining their knowledge of how immune-system memory works and how flu strains evolve, scientists were able to come up with a strategy for vaccination.

A *vaccine* is a substance introduced into your body to trick it into bringing forth an immune response. This response creates antibodies, which can then protect you from the microorganism the vaccine is designed to imitate. Typically, the vaccine is developed in some way from the organism from which immunity is desired. The vaccine can be made from the whole organism or from some of the external parts, because those are the parts that your immune system recognizes. Moreover, vaccines can consist of live or dead organisms.

When you get vaccinated against the flu, you're actually receiving three different vaccines at the same time. That's because three major lineages of the flu are currently in circulation in the human population: two different types of influenza A and one influenza B strain. Because these three kinds of viruses are different enough that no single vaccine can protect against all of them, one strain from each of the three viral groups is chosen for vaccine production.

## Dead vaccines

Initially, a dead vaccine appears to be the better vaccination option, if you consider that the vaccine consists of disease agents that you're going to be introducing into your body. But producing a dead flu-virus vaccine has a major disadvantage: You need to make a lot more of it than you do of a vaccine based on a living virus.

The problem doesn't sound so critical; surely manufacturers can just make more. But in fact, the production of the annual flu virus is a major undertaking. The viruses created for vaccine production are grown in chicken eggs, where the flu happily multiplies. But think about all the eggs you need to grow enough flu vaccine for all the people who need to be inoculated. In the United States alone, you need more than one quarter of a billion eggs. In addition to requiring a lot of eggs, this process takes a lot of time. It's approximately half a year between the time when flu-vaccine production starts and when the first doses are delivered for distribution.

Further complicating this long period of production is the fact that the decision about which three influenza strains should be used to make the vaccine has to be made months before the beginning of the flu season. Because the flu is always evolving, the earlier the decision is made, the less likely it is to be correct. Thus, as you may have noticed, flu shots work better in some years than in others.

Cell-based flu vaccines are newest thing in the manufacturing of flu vaccines. Rather than incubate the virus in eggs, the cell-based techniques grow vaccines in laboratory cell cultures. One of the main advantages of this technique is that, by not being dependent on a huge supply of eggs, the makers don't have the same quantity limitations or contamination challenges.

## Live vaccines

The advantage of using a live flu virus in a vaccine is that a little goes a long way. The virus replicates in a person's body; the immune system ramps up to defeat the infection; and the result is immunity against further infection. Unfortunately, one of the risks is that the live virus is going to reproduce in your body before your immune system knocks it out. That situation complicates the vaccine-production process, because vaccine designers have to create a strain of the virus that won't make you sick but is still similar enough to the more virulent virus to kick your immune system into action.

### Using evolution to create less harmful viruses for vaccines

The flu's ability to exchange genome segments with other influenza viruses has been nothing but trouble for humans. But scientists have turned the tables, using this mechanism to fight back. They've discovered a clever way to create new virus strains that are disguised on the outside with all the bits of the virus humans want to be immune to, but filled on the inside with a collection of different alleles (slightly different forms of the flu genes) that aren't that good at making people sick. Voilà! This "cream-filled" virus reproduces in the body well enough to cause the immune system to attack it, but not so well that the person actually gets ill.

The live-vaccine virus will begin evolving to attack you more efficiently as soon as it starts reproducing in your body. The plan, of course, is that long before enough time has gone by for the virus to develop greater efficiency at reproducing in you, your immune system will have had a chance to defeat it. That's the plan, anyway.

### Using evolution to keep vaccines safe

The fly in the live-virus vaccine ointment is evolution, of course. Even though the live virus used for vaccine production has been constructed so that it's weak and can't reproduce rapidly in the human body, it may not stay that way. As the weakened live virus reproduces, it will produce offspring with random mutations, some of which may improve their ability to live in a person. These progeny in turn will produce offspring, and some of them may be even better at beating the body's defenses. If enough time goes by, a reasonably harmless virus designed for use as a vaccine could end up as one that's much better at eating human cells for lunch.

So can scientists do anything to reduce the chance that evolving viruses will get the best of us? Short answer: Yes, by using the evolutionary process to our advantage.

The influenza that makes you sick is happiest at your normal body temperature. That shouldn't be too surprising, because it lives in the human body, and selection has had a lot of time to tune the flu for 98.6 degrees Fahrenheit. That's one of the reasons your body turns up the heat when you get sick: It's trying to burn the virus out.

Scientists have used this bit of knowledge to their (and your) advantage: They decided to turn down the heat in the laboratory. By growing the flu in colder temperatures in the laboratory, they evolved a flu strain that reproduces well at lower temperatures. These new, cold-adapted flu strains aren't very good at growing at human body temperature.

Why did scientists want this cold-adapted flu strain? Some very clever vaccine designers realized that it's colder in your nose than it is in your lungs. Your lungs are deep inside your toasty-warm body, but your nose is right up there on the edge, breathing air that's generally below 98.6 degrees — especially during the flu season.

By creating a vaccine virus that can reproduce at temperatures slightly *below* your body temperature but not very well *at* your body temperature, the designers developed a very nifty live vaccine indeed. You can spritz a little bit of it in your nose, and although it will reproduce and stimulate your immune system to respond, it can't get from your nose into your lower respiratory system, where it's just too hot. (That's why when you get the live flu virus, the nurse spritzes it into your nose.)

The power of evolution is such that, given enough time, even this virus can get out of your nose. With each round of replication, variants that are a little bit better at growing in high-temperature cells may be produced by mutation. For this reason, people with weakened immune systems, including very young children and elderly individuals, should not get the live-virus vaccine.

## Predicting the future to make next year's vaccine

It's a shame that each year, designers have to redesign the flu vaccine to account for recent flu evolution. It's even more of a shame that getting the vaccine right is so hard. The health officials who make these predictions use a variety of information concerning past patterns of flu outbreaks, as well as which strains are sweeping through different parts of the world. Given the rapid evolution and incredible variation of the flu population at any given time, it's amazing that vaccine designers do as good a job as they do.

Understanding the details of flu evolution can help scientists predict what flu strain may be coming next. We don't know what's going to happen next year (that flu isn't here yet), but we do know what has happened in the past (samples of those flu viruses are in the freezer, and their genetic sequences are known). Scientists can use this information about past years' flu strains to reconstruct the history of flu evolution. Phylogenetic analysis lets them make a family tree for the past years flu strains (refer to Chapter 9 for information on how to construct these trees). This tree is scientists' and virologists' best estimate of what begat what in the never ending progression of new flu strains.

Robin Bush, Walter Fitch, and colleagues used the sequence of the H gene to create such a tree for the H3N2 group of influenza A currently infecting humans. They used data from 1984 through 1996, and they were able to identify which flu lineages in each year gave rise to the flu strains that made people sick the next year.

They then set out to figure out what was it about those strains that predisposed them to being the ancestors of the next big wave of infection. There are lots of little branches of the flu family tree around at any one time — what we want to know is from which of those little branches will the strain that's going to sweep through your school district arise. We want to make a vaccine against that particular sort of flu strain. The trick is figuring out which strain it is.

By figuring our how the flu is evolving in response to our immune system, we learn clues that might help us predict what next years will look like and fine tune vaccine design in advance.

### Making and testing predictions

With 12 years of data in this study, Bush and the others had 11 separate pairs of data sets to test different predictions. First, they looked really hard at the flu RNA sequences for 1985 and asked what they could measure about the 1985 flu strains that could be used to predict the major 1986 flu strain. They then went back to the 1986 records to check to see whether their predictions were right. Then they did the same thing ten more times to test their ideas further. The following section explains what they discovered.

### Anticipating next year's flu: The clues

Possible factors indicating that a particular flu strain is likely to give rise to next year's outbreak include the changeability of a particular strain (maybe the strain that changes the most is the one that spawns the following year's flu) and the changeability at a particular region of the virus (maybe a virus that changes in a particular way is the culprit responsible for next year's flu).

What Bush and colleagues discovered is that the first one wasn't the key, but the second one was:

### The degree of general change in a particular strain.

Maybe the best candidate is the flu strain that's changing the most overall. Because this strain is changing at a faster rate, maybe it will be the one that gets around your immunity to last year's flu.

Sounds good, but this predictor didn't turn out to be an especially accurate one. Having lots of random mutations is no guarantee that those mutations result in genes that are better at fooling our immune system.

### The degree of change at particular regions.

Maybe the flu strain most likely to cause next winter's problems is the strain that's changing in some particular way — changing the most at just the regions where the H protein is interacting with the human immune system, for example. This strain would be more likely to catch your immune system napping!

Turns out we don't know the exact details about which parts of the flu are the most important in interacting with our immune system. But by studying the flu's evolution we can find out! By looking at the flu stains that had been successful in the past the researches were able to identify specific sites in the H gene where changes were associated with this success.

Bush and coworkers discovered that the more mutations that occurred in a particular isolate at just these key locations, the more likely that isolate was to spawn the next year's flu. These are the flu strains that seem to be rolling with the punches of the human immune system, and they're ones we need to be prepared for in the future.

## Making a more universal influenza A vaccine

Because the H and N proteins that cover the surface of the flu are so variable, designers have to reassess which particular strains they should use as models for each year's vaccine (refer to the preceding section). It's a shame that the surface of the flu is so variable from one viral strain to the next; things would be a lot more convenient if they were all the same.

Fortunately, one part of the surface of influenza A is very uniform across many isolates: the Matrix, or M, protein. When your immune system finds the flu in your body, it doesn't seem to pay much attention to the M protein, as H and N proteins are much more obvious targets. But that protein is there, and it's possible to make your immune system see it, too.

Walter Fiers and colleagues set out to do just that. They manufactured virus-sized particles covered with the M protein. Once Fiers and company had the little M-protein-covered particles, they tested them on mice to see whether they could stimulate an immune response that would protect the mice from influenza A. They found that even though the mice still got sick, they got much less sick than the mice that had not been vaccinated with these particles.

The fact that it is possible to make the immune system respond to the M protein raises a question: How would evolution of the influenza A M protein be affected by the presence of many such vaccinated individuals? The M protein is very uniform among different influenza A strains right now. If many people had antibodies that attached to the M protein, however, that might select for increased variation in the M protein. All the influenza A viruses make their M proteins in about the same way, but that doesn't mean they have to. Perhaps the viruses can make the M protein in many ways.

Walter Gerhard and colleagues set up an experiment to see how influenza A viruses evolved in the presence of antibodies that targeted the M protein. They took mice with weakened immune systems and gave them antibodies against the M protein. They infected the mice with an influenza A strain, and after several weeks, they examined the descendents of the original viruses to look for changes in the M protein.

Because these mice had weakened systems, they were not able to eliminate the flu virus rapidly. Because the flu virus was replicating in the mice, random mutations were occurring, and it turned out that some of these mutations did result in changes in the M protein. The scientists found only two new variants of the M protein, however, rather than the huge number of variants that would occur in the H protein in the same sort of experiment.

These experiments are encouraging, because the results suggest that the influenza particle may have only a limited number of ways to make a functional M protein. With any luck, it might be possible to vaccinate against all influenza A viruses at the same time using a vaccine based on M protein.

This type of experiment shows the potential of incorporating evolutionary information in drug design. By trying to target a part of the flu that seems more evolutionarily constrained, it may be possible to create a more universal vaccine for use in the event of a dangerous outbreak. In the case of an extremely pathogenic flu, a vaccine that offered only partial protection would be much better than no protection at all. A vaccine like this could provide one line of defense against the threat of the H5N1 avian influenza.

In fact, British and American researchers have recently tested such a vaccine (one based on the M protein), and the results are very encouraging. They indicate that the vaccine could provide universal protection against the Influenza A virus.

# Part V
# The Part of Tens

The 5th Wave                    By Rich Tennant

## In this part . . .

C'mon. You can't possibly expect to get through a book on evolution without at least a little information on fossils. So this part has a list of ten of my own personal favorites, as well as a list of adaptations that I'm particularly fond of.

And let's be honest; you were probably also expecting a little bit of information about what the evolution naysayers have to say (beyond "nay"), as well as some discussion on what intelligent design is all about. As hesitant as I am to include non-science-parading-as-science in this book, I have included a list of ten arguments you'll hear against evolution. And then I explain why those arguments don't hold water.

# Chapter 20

# Ten Fascinating Fossil Finds

*Y*ou can't talk about evolutionary biology without talking about fossils. They're the only physical evidence of plant, animal, and microbial forms that existed long before the first primordial human.

In this chapter, I offer ten fossil finds. I chose these fossils because they reveal information not only about specific organisms (such as miniature mammoths), but also about evolution and the world. I didn't organize this chapter in any particular order, but really, how can you begin a discussion of fossils and not begin with dinosaurs? You can't. So I start there.

## Dinosaurs

Since the early 1800s, when the first dinosaur fossil was discovered (and recognized for what it was), people's imaginations have been fired by dinosaurs — not only because of their size (they included some of the largest animals the world has known), but also because they seem so unlike animals today.

What does the fossil record say about these beasts? Quite a bit, actually, but not nearly as much as you'd think, despite the attributes given to them by movie directors and novelists:

✔ They lived from approximately 230 million to 65 million years ago — a period long enough for them to see some species go extinct and other species arise.

✔ They were not lizards, even though the word *dinosaur* comes from the Greek, meaning "terrible lizard" and they have a common ancestor with lizards. The skeletal differences between lizards and dinosaurs are pronounced enough that dinosaurs get their own branch of the tree of life.

Dinosaurs' legs, for example, were directly under the body rather than off to the side, as is the case with lizards and crocodilians.

✔ They didn't live at the same time as humans. Modern humans have been here for only a few hundred thousand years, and the dinosaurs went extinct (mostly) 65 million years ago.

✔ No one can claim to know what color dinosaurs were (the fossil record doesn't give information about color), though it's reasonable to assume that, like every other group of animals, they came in a variety of colors.

✔ The largest dinosaurs may have weighed 100 tons or more. Why the uncertainty? Many dinosaur fossils are incomplete. It's reasonably straightforward to come up with a good estimate of an animal's weight based just on its skeleton — if you have a complete skeleton. The largest dinosaur species for which scientists have found all the various bits is Brachiosaurus, which weighed in at about 30 to 40 tons, but bones from much larger species have been found. Unfortunately, without all the bones, paleontologists can't be sure exactly what those dinosaurs looked like.

✔ Some species had social behavior. Fossilized tracks indicate that large groups of dinosaurs traveled together. Nesting areas where several dinosaurs made nests together have also been found. Evidence even exists that baby dinosaurs remained in the nest and were cared for by the parents (because fossils of young but not newly hatched dinosaurs have been found in nests).

✔ No one knows for sure what caused the dinosaurs to go extinct. One theory is that a large meteor smashed into the Earth, raising a cloud of dust big enough to alter sunlight and weather around the globe. As it turns out, scientists have found evidence that a large meteor struck the Earth around the time when the dinosaurs went extinct. The geological boundary corresponding to the time when the dinosaurs disappeared contains a layer of iridium — an element that is rare on Earth's surface but plentiful in some meteors; it would have settled on the Earth as dust from the sky after a meteor impact.

✔ One group of dinosaurs is still around today. You know this group as birds. The bird lineage has its origin within the dinosaur branch of the tree of life — an initially controversial idea that's now widely accepted.

The first "scientific" discovery of dinosaur fossils occurred in the early 1800s, but it's surmised that ancient peoples (such as the Romans and Chinese) discovered dinosaur fossils earlier. Those fossils could be the source of myths about dragons and other fantastical beasts that crop up in lore and literature from ancient times.

# Archaeopteryx

Archaeopteryx (see Figure 20-1) is an example of what is often referred to as a missing link. Except of course that it isn't missing anymore because we found some! Archaeopteryx is a species of bird-like dinosaur that has many of the characteristics of modern birds — feathers, wings, flying — stuff like that. But it also has characteristics typical of the dinosaur lineage from which it arose: most obviously its toothy grin!

Archaeopteryx was discovered only a couple of years after Darwin published his book *On the Origin of Species.* As he said in his 4th edition:

> Until quite recently these authors might have maintained, and some have maintained, that the whole class of birds came suddenly into existence during the eocene period; but now we know, on the authority of Professor Owen, that a bird certainly lived during the deposition of the upper greensand; and still more recently, that strange bird, the Archeopteryx, with a long lizard-like tail, bearing a pair of feathers on each joint, and with its wings furnished with two free claws, has been discovered in the oolitic slates of Solenhofen. Hardly any recent discovery shows more forcibly than this how little we as yet know of the former inhabitants of the world.

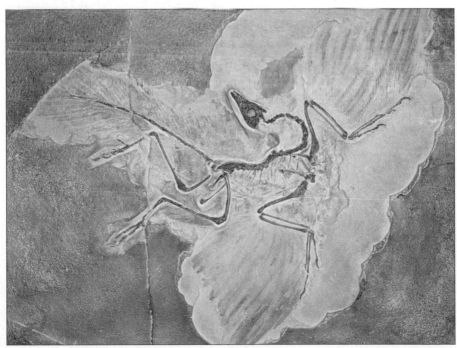

**Figure 20-1:**
Archae-
opteryx.

© Naturfoto Honal/Corbis

As you can imagine, this discovery caused quite a stir for such a small creature. Archaeopteryx was about a foot and a half from nose to tail, hardly what we think about when we think about dinosaurs, but — ignoring the distraction of wings and feathers for a minute — many of its structures are clearly related to the dinosaurs. So it turns out the dinosaur branch of the family tree did not entirely die out. One small branch, the birds, remains with us still.

That birds might be a surviving lineage of the dinosaur branch of the tree of life was originally proposed by one of Darwin's contemporaries, Thomas Huxley, but the idea never really caught on. It wasn't until the 1970s and the developments of phylogenetic analysis that the question was revisited and found to have overwhelming support.

# Wrangle Island Mammoths

Woolly mammoths lived at the same time as early humans but went extinct about 10,000 years ago. At least scientists thought so until the late 20th century discovery of fossils on Wrangle Island, off the coast of Siberia. Mammoths were living on this island as recently as 4,000 years ago — the same period when the Egyptians were building the pyramids.

These island mammoths are especially cool not because they survived to historical times (though that fact is fascinating), but because they were miniature mammoths, only about 7 feet tall — the Shetland ponies of the mammoth family. Why would pint-size mammoths have evolved? Possibly because there was no advantage to being huge on islands.

One of the major advantages of being a very large animal is that you're just too big for any predators to attack. When an elephant — or a mammoth, I suppose — reaches a certain size, nothing is going to eat it. But being large also has costs. An African elephant needs to eat around 300 pounds of food and drink 50 gallons of water every day. The evolution of smaller mammoths could have been selectively advantageous if the species had no predators on the island. Smaller mammoths that could survive on less food would be less likely to starve.

# Pterosaurs (Pterodactyls)

The largest living flying bird in the world is the Andean condor, which can have a wingspan of more than 10 feet. Go back in time, and you can find fossils of an extinct bird with a wingspan twice as big — a creature that would make Big Bird look small. To find the largest animal that ever flew, you need to go back even farther — 65 million years back, in fact, to the time of the pterosaurs (or pterodactyls, as they are often called).

Pterosaurs were flying reptiles related to dinosaurs. Although some of them were quite small, the largest had a wingspan that reached almost 40 feet. Perch a pterosaur atop a school bus, and its wings would stretch the entire length of the bus. (Hang on to your children!)

Evolutionists like pterosaurs not just because the group includes the largest creature that ever flew, but also because it's represents yet another independent origin of the ability to fly. Pterosaurs' wings were structured differently from those of other vertebrate lineages that have taken to the air: birds and bats (which also differ from each other). In bats, for example, the wing is supported by the elongated digits of the forelimbs, but in pterosaurs, the wing was supported by an elongated third finger with the other digits being much reduced. Both mechanisms work to make a wing, but the random process of mutation and differences in initial structures nudged these lineages down different pathways.

# Trilobites

If you ever owned an animal fossil, chances are that it was a trilobite — one of the most common types of animal fossils (see Figure 20 2). Trilobites were arthropods (things like lobsters, spiders, and bumblebees) that lived in the ocean. They appeared in the fossil record more than 540 million years ago and went extinct 250 million years ago.

The fossil record of trilobites is especially rich because aquatic sediments provide the perfect environment for fossilization. In addition to the trilobite fossils themselves, scientists have found fossils showing the organisms' burrows and tracks. Here's what these fossils tell researchers:

✔ Different trilobite species evolved different body shapes. Some were smooth; others were extremely spiny. They ranged in size from a few millimeters to a couple of feet.

✔ Some lineages had the ability to dig burrows; others were able to roll up in balls, like pillbugs and hedgehogs; and some evolved the ability to swim.

✔ They inhabited shallow waters as well as deep waters, though they have not been found in any freshwater environments.

**Figure 20-2:**
A trilobite.

Just as important as the trilobite-specific information that these fossils offer is what they show about the time frame of evolutionary transitions in lineages with a plentiful fossil record. We have so few human fossils, for example, that it's reasonable to assume we will miss transitional forms because the hominid fossil record is poor.

Even though trilobite fossils are numerous, we still see sudden (in geological time) transitions and the rapid appearance of new forms in this lineage. Where speciation events occur in a small, localized area, the probability of finding intermediate forms decreases, and the fossil record appears to contain more sudden transitions as species appear and go extinct. Head to Chapter 8 for info on speciation.

## Tiktaalik Rosea

Perhaps you've seen cartoons depicting the transition of vertebrates from water to land animals. Usually, these drawings show some sort of fish with stubby little legs, which isn't too far off the truth. Scientists recently found a fossil quite similar to the type that's famous from all those evolution cartoons: Tiktaalik rosea.

Tiktaalik has a lot of fishlike structures, such as fins, scales, and gills. But its "fins" have a leg-like structure with wrists and parts like a hand or fingers (see Figure 20-3). This critter could do pushups! It's not clear to what extent Tiktaalik left the water, if it did at all. Some scientists hypothesize that this new fin structure would have been advantageous for maintaining position in fast-flowing shallow water.

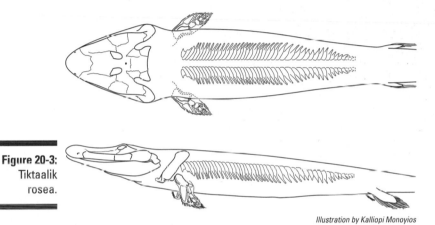

Figure 20-3: Tiktaalik rosea.

*Illustration by Kalliopi Monoyios*

Other interesting characteristics of this fossil are its neck and the eyes on top of its head — more like a crocodile than a salmon. What scientists have discovered from this fossil is that the anatomical modifications associated with the transition to land may have evolved first in the water.

# Hallucigenia and the Burgess Shale

The Burgess Shale is a concentration of fossils in the Canadian Rockies containing large numbers of morphologically distinct species that at first glance don't seem much like the organisms that are around today.

The fossils found in this shale are cool for two reasons:

✔ Their excellent degree of preservation often reveals intricate structures of the softer body parts.

✔ The large diversity of forms that were new to science raises interesting questions about the evolutionary history of life on Earth.

For the most part with fossils, it's possible to identify the major group to which the organism belongs. For example, you might know that a particular fossil is a clam, perhaps not like any clam we have today, but a clam nevertheless. Before the discovery of the Burgess shale, scientists could identify fossil animals as earlier members of particular branches of the tree of life. Trilobites (see the preceding section), for example, were recognizable as members of the arthropods branch.

But when the animals in the Burgess Shale were examined, it wasn't clear immediately what groups many organisms belonged to, and in some cases, such as the organism hallucigenia, it wasn't even immediately clear which end was up (see Figure 20-4). Creatures like hallucigenia make one think that perhaps many such interesting and unusual creatures have evolved and subsequently gone extinct.

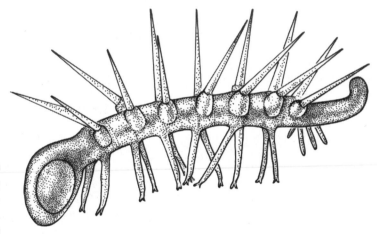

**Figure 20-4:**
Halluci-
genia.

*Drawing by Mary Parrish. Courtesy of Smithsonian Institution*

On subsequent analysis, and with the passing of considerable time and effort, scientists have been able to assign many of the fossils found in the Burgess Shale to existing groups. But some fossils still defy placement and may indeed be examples of unique body plans that turned out to be evolutionary dead ends.

## Stromatolites

Stromatolite fossils dating back more than 3 billion years are among the oldest fossilized evidence of life on Earth. Stromatolites are pillars formed by layer upon layer of microorganisms, such as cyanobacteria (see "Micro-fossils," later in this chapter). These layers trap sediments that become incorporated into the structure, or in some cases, the organisms themselves are responsible for producing calcium carbonate, which becomes a component of a structure. Either way, when the structure is fossilized, the layers are visible as bands in the columnar structure.

Stromatolites are common in the fossil record; however, it wasn't until the discovery of living stromatolites that we could be sure that these structures were the result of biological activity. While there certainly seem to have been a fair number of stromatolites formed billions of years ago, they are a lot less frequent now. Living stromatolite structures are currently found only in areas where there are no predators to feed on the layers of microorganisms. In the days before anything else had evolved to eat them, being eaten wasn't a problem. Today, they're just out of luck in most places.

# Microfossils

Microfossils are simply fossils too small to study with the naked eye. Micro-fossils are grouped based solely on their size and don't correspond to a particular branch of the tree of life. Following are a few examples:

- ✔ **Cyanobacteria:** The earliest cyanobacterial fossils date from approximately 3.6 billion years ago, and for the longest time, such single-cell organisms were the only life on the planet. Interestingly, the earliest fossils of these microorganisms are extremely similar to those that exist today.

- ✔ **Foraminifera:** Organisms typically about a millimeter in size that produce hard shells, with each species being unique. Because the species present at any given time vary (as a result of extinction and speciation), scientists can use the species to date different geological strata. If they find the same collection of foraminifera species in two samples, they know that the samples date from the same time.

  That info may seem rather useless, but it's pretty handy for marine oil exploration. Suppose that you find oil in strata of a particular age. Naturally, you want to explore the strata in a nearby location. But because of geological activity, the strata may be at a different depth in the second location. By analyzing the various species of foraminifera across the two locations, you can match strata of similar ages.

- ✔ **Spores and pollen:** Researchers can use changes in the composition of plant reproductive structures over time to glean information about how communities change through time. Scientists can use pollen samples to make inferences about climate change over shorter time scales, observing how pollen from plants characteristic of one set of environmental conditions is replaced in subsequent strata by pollen from plants characteristic of a different set of environmental conditions.

# Amber

Amber is fossilized tree sap. It's quite commonly used in jewelry and has a beautiful golden color. As nice as the sap itself is, the really interesting thing about amber is that organisms — plant parts (such as leaves, seeds, and flowers), small mushrooms (representing fungi), insects, snails, spiders, and even frogs — can become trapped in it. These trapped organisms retain incredibly fine detail. Flowers, for example, can appear perfectly preserved. For groups that are poorly represented in other forms of fossils, amber provides the best clues about the timing of various evolutionary events.

Spiders, for example, are quite rare in the fossil record because they don't have hard exoskeletons, but these creatures, and even their fossilized webs, turn up in amber. Data from amber allows researchers to date the evolution of spider webs to more than 100 million years ago. A 130-million-year-old spider-web fossil has been preserved with enough detail that you can see the individual droplets of glue the spider placed on the web!

# Chapter 21

# Ten Amazing Adaptations

**A**daptations are evolutionary changes resulting from natural selection. The environment in which a species lives exerts selective pressures on species such that they change over time.

This chapter lists ten adaptations I particularly like because they represent the types of adaptations that are possible. Some are no-brainers. Want to live in the ocean, and you're a mammal? You'd better be streamlined and insulated. Others are breakthroughs. Photosynthesis, for example, changed the chemistry of the entire world because of all the oxygen it produced. And some are — almost — out of this world, such as the creatures that have evolved to live in the deep-sea thermal vents that you see on the National Geographic Channel.

# Different Kinds of Teeth

You have different kinds of teeth, but have you ever thought about how helpful those differences are? According to the fossil record, the ancestral condition in reptiles and the reptilian lineage that led to mammals is to have teeth that are all the same. Think about a crocodile; it has lots of teeth, but they're all the same. The same goes for the dinosaurs.

But then things changed. The Dimetredon, a dinosaur-like creature that had a sail on its back, had two kinds of teeth. Having two kinds of teeth allows for the possibility of division of labor. The teeth in the rear can be used for processing food — for grinding in grazing mammals or for slicing up chunks of flesh in your house cat — and the teeth in the front can specialize in food acquisition or other functions: snipping plants in herbivores, subduing prey in carnivores, or other functions such as defense.

What's interesting about the evolution of teeth shape is that the lower jaw has to match the upper jaw. As a consequence, changes to both jaws have to occur at the same time. For this reason, teeth change very slowly. Things get a little more complicated for humans. We've taken to capturing and processing most of our food outside our mouths. We cook it (ever try to eat raw rice?) and use knives and forks to cut it (ever ripped raw meat apart with your teeth?). As a result, our dentition is much reduced.

Evolution has also gone in the other direction: an ancestor with different kinds of teeth evolving into a species with uniform teeth. Example are animals that don't need to process their food because they swallow it whole — like dolphins and some other marine mammals. These creatures are descended from ancestors with non-uniform teeth, but they evolved dentition with one kind of tooth designed for grabbing small, slippery prey, which they then swallow whole.

# Sight: The Evolution of the Eye

Eyes are beautifully complex structures, and their evolution was a source of some mystery to Charles Darwin. The idea that the eye could not have arisen from the process of natural selection is a common misconception even today and is rooted in the idea of *irreproducible complexity* (see Chapter 22), which states that complex structures could not have arisen as a result of a gradual evolutionary process because humans can't imagine how intermediate forms would be advantageous. You often see this argument stated this way: "What good is half an eye?"

Darwin never suggested that natural selection couldn't produce the eye, of course; he just admitted that he didn't know exactly how the process unfolded. He imagined that many intermediate steps had to occur, leading from a very simple light-sensing structure to the structures you're using to read this page; he just didn't know what they were. Fast-forward to today, when scientists know that many of the intermediate stages exist in other animals. From this fact, they can imagine the series of small steps that would lead from the simplest light-sensitive cell to a more complex eye. For example:

- **Step 1: Start with the simplest light-sensitive cells.** A patch of these cells can determine the presence or absence of light but not much else.

- **Step 2: The cells are set slightly into the body in a little pit or cup.** After the cells reside in a depression in the surface of the organism, the light-sensing apparatus becomes capable of determining the direction from which the light is coming.

- **Step 3: The edges of the pit grow together so that light enters the pit through a very small opening.** This arrangement is the principle behind a pinhole camera. Even though the camera has no lens, restricting light to traveling through a small hole results in a crisper image.

The principle behind the pinhole camera is simple physics, and you can test it yourself. If you have to hold a book at a distance to focus (you young folks wait a few years), poke a tiny hole in a piece of paper and peer through it. You can read the text without having to move the book away (or so far away).

✔ **Step 4: A lens is added to the opening of the light-sensitive cells.** You don't have to imagine the lens evolving all at once as a lens; you can easily imagine that a layer of translucent cells over the opening of the pinhole had a protective function. And we have plenty of examples of see-through cells in the animal kingdom. When that layer was in place, any changes that resulted in a crisper image would be selectively advantageous.

Imagining such intermediate steps goes a long way toward helping you see how a series of small changes can lead to complex structures like the eye.

# Cave Blindness

A common pattern that's repeated across a large number of animals in different locations is the evolution of *cave blindness* — the evolution of sightlessness in lineages that have come to inhabit caves. Cave blindness is an excellent example of convergent evolution, in which the same trait evolves independently in different organisms (refer to Chapter 9).

The ancestors of most cave-dwelling organisms came from non-cave environments that had light. In fact, you can go to any big cave with its own ecosystem full of cave critters, and you'll find blind cave animals whose closest relatives (in the tree of life) can see.

You can easily see why the selective pressures on organisms existing in darkness would be different from those existing in light: Perhaps the energy required to produce those structures was needed for other functions — it's wasted making eyes in the dark and is better spend it some other way. Perhaps rewiring the sensory system — cave critters often have a good sense of smell, extra-sensitive tactile feelers and antenna, or other stuff that's good to have in the dark — requires minimizing the eyes.

A common theme in evolutionary biology is that things can happen over and over. If something is good once, maybe it's good twice. Cave blindness is an example; over and over, organisms that move into caves lose their sight. Flight (discussed in "Vertebrate Flight," later in this chapter) is another good example; it evolved in insects, mammals, birds, and even the extinct lineage pterosaurs (which you can read about in Chapter 20).

Finally, as discussed in Chapter 5, we always need to be cautious about assuming that an evolutionary change is an adaptation. Once it's dark, mutations that degrade the eye are no longer bad; they just don't matter. So over enough time, genetic drift might be expected to result in the loss of eyes even if the change isn't adaptive — in which case cave blindness is an example of evolution but not adaptation.

# Back to the Sea

Eons ago, the first vertebrates crawled out of the ocean and onto land. It seemed like a good idea at the time. But since then, a few species have headed back to sea. Some have gone back completely; others return to the land to reproduce. Sea snakes, penguins, seals, whales, manatees, and sea otters are examples of animals that evolved independently back to sea animals from terrestrial ancestors.

Through phylogenetic analysis (see Chapter 9), scientists can tell that vertebrates have returned to the sea on several occasions; that in each case, the aquatic group is nested within a larger terrestrial group; and that the common ancestor of all the individuals in the group was terrestrial. DNA-sequence evidence has been especially helpful in confirming relationships. The skeletal structures of whales and hippos, for example, provide evidence that they're related, even though they don't look all that similar. Through DNA evidence, scientists also know that the seal group is nested within the carnivores.

Living in the ocean selects for several suites of characters:

- **With the exception of the sea snakes, which live in warm tropical waters, all these groups have made adaptations to keep warm.** Penguins have an insulating layer of blubber, as do all the mammals that live in the ocean, except for sea otters, who have water-resistant coats that trap insulating air layers.

- **All ocean-dwelling species are reasonably streamlined to facilitate faster motion in water.** The manatee group is the least streamlined and, not surprisingly, the slowest. But then, manatees are also vegetarians, and plants don't run very fast.

- **Their appendages have become modified for locomotion in water:** flippers in the case of many of the mammals; wings that act like flippers and webbed feet in the case of the penguins. Sea snakes have a flattened body, especially in the tail region, allowing them to swim with an eel-like motion.

- **All have characteristics that indicate their descent from land animals.** They all breathe air, for example, which can be very inconvenient when you live in the water, but each has evolved a rather impressive capability for holding its breath. (Sperm whales can routinely hold their breath for an hour!) They also have vestigial structures indicative of terrestrial ancestors. One of the best examples is the hind leg bones of whales. These small bones are completely within the body (whales don't have hind limbs) but correspond to the rear leg bones of terrestrial quadrupeds.

# And Back to the Land Again

Current research suggests that the elephant may have an aquatic ancestor. Keep in mind, though, that the jury's still out on this particular hypothesis. But I decided to share it with you nonetheless because it's a good story. Think of it as a trip to the cutting edge of science but remember: There's not a lot of room on the edge, so sometimes we fall off!

The evidence that the elephant evolved from an aquatic ancestor is threefold:

- **Elephants' closest living relatives are manatees.** The fossil record for the evolutionary transition of the manatee lineage to the aquatic environment is reasonably complete. Fossil manatees with the vestigial legs have been found, and in Jamaica recently, scientists discovered a fossil manatee with functional legs that could support the weight of the animal yet still showed adaptation to an aquatic lifestyle. The aquatic ancestor of elephants would have been similar to such a creature, but the elephant lineage returned to the land rather than transition to a fully aquatic lifestyle, as the manatee lineage did. What we know from the fossil record and the structure of existing manatees and whales is consistent with the hypothesis that the most recent common ancestor of these beasts already had its feet in the water. One branch of the family tree kept going, while the other headed back to shore.

- **Elephants are surprisingly good swimmers.** Although they can't raise their heads out of the water while swimming, they can use their trunks as snorkels. If you've ever breathed through a snorkel while standing in water (rather than floating on the top of the water), you know that this feat is difficult because of the pressure of the water. Well, snorkeling is even harder for elephants because of the length of their trunks, but they do it relatively easily because they don't have a *pleural cavity* (a membrane surrounding the lungs that's unique to mammals). The absence of this cavity prevents the pressure difference from damaging the lungs.

- **Elephants have traits that are common to mammals that have returned to the ocean.** Chief among these traits is internal testes — a characteristic found in no other land mammal but in all aquatic mammals.

# Photosynthesis

Photosynthetic organisms convert light energy from the sun to chemical energy, which they use to power their bodies. Photosynthesis is responsible for almost all the energy used by organisms on this planet. Either directly, as a result of internal photosynthesis, or indirectly, by eating something that photosynthesizes itself (or that ate something that did), most species run on photosynthetic energy. Because oil is ancient fossilized plant matter, your car is running indirectly on photosynthetic energy, too.

When you think of photosynthesis, plants are probably the first (and likely the only) organisms that come to mind, but they aren't the only organisms that use photosynthesis. Some bacteria do, too. Figuring out the origin and evolution of the various chemical mechanisms by which bacteria photosynthesize is a source of active research. As with so many things, biologists are still learning about the evolutionary diversity of photosynthetic mechanisms. In the summer of 2007, a new type of photosynthetic bacteria with different chemical pathways was discovered in a hot spring in Yellowstone National Park.

# Deep-Sea Thermal-Vent Organisms

Photosynthesis may be responsible for most of the energy that organisms use, but a group of organisms that live deep in the ocean use another source of energy. Down at the ocean bottom, in regions of sub-oceanic volcanic activity, are thermal vents that spew out hot, mineral-rich streams of water from within the Earth. These mineral-rich streams of water, which were unknown before the 1970s, can be used to generate energy in much the same way that photosynthesis does.

Specifically, some bacteria can generate energy by oxidizing the hydrogen sulfide (the substance that makes rotten eggs smell rotten) that is present in the hot vent water. So these organisms derive energy not from photosynthesis, but from *chemosynthesis.* The energy that forms the base of the food chain in these deep thermal vents is not the energy of the sun, but the energy at the core of the Earth.

Just as plants form the basis of an entire community on the surface of the earth, these bacteria form the basis of a whole community on the ocean floor. And what a community it is! Creatures found nowhere else live near these thermal vents. The sulfur-oxidizing bacteria live within large worms that have no digestive systems but derive their energy solely from the bacteria, which in turn get a secure location anchored right near the stream of hydrogen sulfide.

Finally, these deep thermal vents have a very dim glow, so additional photosynthetic bacteria may be lurking there somewhere. Things are always more complicated than they seem at first!

# Endosymbiosis

One of the most amazing evolutionary events is called *endosymbiosis*. According to this theory, some of the structures in eukaryotic cells — such as mitochondria and chloroplasts — once were free-living bacteria that became engulfed in ancestral eukaryotic cells, and a symbiotic relationship evolved. Somehow, and we don't completely understand how, two ancient critters joined up and eventually became so tightly interdependent that they effectively became a single organism. Remember that we're eukaryotes so that means that we are derived from two different lineages that came together deep in the distant past to make the eukaryotic cell. That's why your mitochondria have their own genome — their distant ancestors used to fly solo.

Here are the details supporting this theory: Mitochondria and chloroplasts bear a strong resemblance to bacteria. When the eukaryotic cells divide, the organelles divide too, and the division process of these organelles is reminiscent of the division of bacteria. Most importantly, these organelles have their own DNA, and analysis of the DNA sequences shows that the organelles are closely related to some free-living species of bacteria.

As you can imagine, this hypothesis was quite controversial initially. Think about it: Descendents of ancient bacteria are living in all your cells. But the DNA evidence seems to be beyond doubt. Your mitochondria have their own genome, albeit much reduced, and it's a lot more similar to a bacteria genome than it is to anything in your nuclear genome. Luckily, we eukaryotes are all living happily ever after, and we've been doing so for at least 2 billion years.

# Vertebrate Flight

Flight is another trait that has evolved several times in several lineages. Birds, bats, pterodactyls — all evolved true flight, and a couple of others have rudimentary gliding ability. Flight is a remarkably successful trait. Groups that are capable of flying radiate extensively. Bats account for one quarter of all mammal species, for example. And pterodactyls, though extinct, survived for 150 million years and encompassed many species, including the largest creatures ever to take to the air.

Theories about the evolution of flight can be divided into two groups: up from the ground and down from the trees. The second group is easier to visualize, because species living today, such as flying squirrels (which don't really fly, by the way), have structures that allow them to glide from tree to tree. Current thinking is that bats arose from an arboreal ancestor, so the gliding hypothesis may apply to them, too. Pterodactyls, on the other hand, don't seem to have descended from arboreal ancestors, so maybe both mechanisms are possible.

What scientists do know is that bats, pterodactyls, and birds evolved flight structures in different ways. In bats, the wing is constructed of a membrane stretched between what for humans would be the fingers of the hand. In the case of the pterodactyls, the wing is supported by just one elongated digit. In birds, the wing is comprised of feathers all along the arm. And that list covers just the vertebrates; flight has also evolved in the insect lineage. Bottom line: Several different mutational pathways generated wings.

# Trap-jaw Ants

Trap-jaw ants are species of ants in which the mandibles (the jaw-like things that they grab prey or bite you with) are locked open and have a trigger that allows them to spring shut with great force. In one case, the jaws snap shut at speeds reaching up to almost 150 miles an hour. The principle is something like an archer drawing a bow: You pull and pull to load the bow, and when you suddenly release the bowstring, the energy is transferred to the arrow. This adaptation is cool on several levels:

- ✔ The jaw speed — 150 miles per hour — is the fastest attack motion in the animal kingdom.

- ✔ This trait has evolved at least four times in four different groups of ants. And although the final outcome is the same in all cases, the exact pathway by which the trait was obtained varies from one species to the next. Specifically, different parts have been modified to serve as the trigger in different ant species.

- ✔ At least one species uses the great force generated by the snapping jaws for functions other than biting — for example, as a means of escape from predators. To escape, the ant points its head at the ground, releases its jaws, and is propelled rapidly away from whatever it's trying to escape.

# Chapter 22

# Ten Arguments against Evolution and Why They're Wrong

*F*or an idea that's almost beautifully simple, evolution certainly has gar-
nered a lot of bad press. To hear some people talk, you'd think Darwin
himself was the devil incarnate; evolutionary biologists are his handmaids;
and people who teach evolution in the classroom are corrupting the minds of
children across the land. Can we get a little perspective, people?

I wrote this chapter specifically to provide a little perspective — and facts to
arguments that tend to lack them. Here I present the arguments some people
make against evolution and then explain why these arguments are wrong.

## It's Only a Theory

Yes. Evolution is a theory, but not in the way the naysayers mean. When they
say it's only a theory, they mean it's only an idea—a guess, if you will. But as
Chapter 2 explains in quite a bit of detail, a scientific theory is not merely a
best guess. It's a hypothesis that — through experiment after experiment,
study after study, analysis after analysis — has yet to be refuted.

Having said that, evolution is not *only* a theory; it's also a fact. The key to
understanding evolution is to recognize how it can be both:

✔ **As a fact:** Evolution is simply genetic changes occurring through time in a group of individuals (a population, a species, and so on). Scientists *know* that these changes occur. They can see the changes; measure them; and, in many instances, figure out when they happened.

✔ **As a theory:** Evolutionary theory seeks to explain what's responsible for the evolutionary process — in other words, what causes these changes. What scientists know today is that natural selection (Chapter 5) and genetic drift (Chapter 6) are two key forces driving these changes.

# It Violates the Second Law of Thermodynamics

The second law of thermodynamics states that *entropy* — essentially, randomness — increases (or stays constant) in a closed, or isolated, system; it cannot decrease. In other words, left on their own, isolated systems become more uniform, not less. The differences smooth out until one common state exists. Think about a glass of ice water. After the ice goes in, the water gets a little colder, and then the ice melts: The entropy has increased in the glass. According to the second law of thermodynamics, the whole universe is doing the same thing: Increasing entropy is "smoothing out" the world. Rather than having hot regions and cold regions, for example, the world would have all its parts the same temperature.

So what does this law have to do with evolution? Diversification of life on Earth has involved very complex organisms evolving from simple forms that were present a few billion years ago — a fact that seems to fly in the face of the second law, because on Earth entropy is decreasing, not increasing. And there's the key. Earth is *not* a closed system. It gets loads of energy from the sun, and that energy is what powers the increase in complexity.

# It's Been Proved Wrong (by Scientists!)

I love this one! This argument stems from the fact that in the hundred-plus years since Darwin's death, scientists have contributed to his original thoughts and refined their understanding of evolutionary events and principles. The spin you see in lots of articles, though, implies that a particular piece of research is at odds with what Darwin thought and, therefore, is proof that Darwin got things wrong.

The best example is the importance of random factors — genetic drift (see Chapter 6), which is one of the key insights modern evolutionary biologists have added to our current understanding of evolution. What scientists know today that they didn't know during Darwin's time is that random events, as well as natural selection, can be evolutionary forces; that random events can be evolutionarily important is an example of a major change in theory of evolution, but it doesn't negate in any way Darwin's theory of evolution by natural selection. It simply makes his ideas more broadly applicable.

# *It's Completely Random*

How long would it take a million monkeys hammering away on a million typewriters to produce *Moby-Dick?* Who knows? (How long did it take one monkey hammering away on one typewriter to come up with the premise for "Who Wants to Marry a Millionaire?") The point? That a complex work — whether it's a Shakespearean sonnet or a book about evolution — can't possibly be the result of random processes.

The problem is that people who make this argument are confusing the fact that some of the evolutionary process of natural selection involves random events with the idea that the whole process is random. True, the mutations produced are random (that is, not directed toward a goal), but natural selection sorts through these mutations in a nonrandom fashion, selecting for those that increase fitness.

A major stumbling block that prevents many religious people from accepting what science has learned about the evolutionary process is the idea that evolution is connected to a random process, the one whereby DNA sequences are passed inexactly from one generation to the next — in other words, the process of mutation. Yet the very process of replicating the DNA is error prone. Scientists can measure the rate at which errors occur in DNA replication just as they can measure the rate of radioisotope decay, but whether an error occurs in one location or another is random.

The random aspect is unsettling for many people. Although we *know* that, given the amount of time available, the process of natural selection acting on randomly produced mutations is more than sufficient to generate our own species, this viewpoint is at odds with some people's view of humanity's place in God's universe. To reconcile the role of randomness with the religious beliefs that things happen for a reason or with purpose, some people suggest that nothing is truly random — that perhaps God set into motion the series of events that caused exactly the particular sequence of mutations that resulted in *Homo sapiens.* Maybe. But no way exists to scientifically measure whether God is or isn't directing these mutations. So these possibilities are outside the realm of science.

# It Can't Result in Big Changes

According to this argument, some changes (namely, the small ones, a mutated nucleotide here or there) can be the result of the evolutionary process, but others (namely, the big ones) can't be. The key areas of dissention are

- ✔ **Speciation:** The argument goes like this: Although evolution can lead to changes within a lineage, it can't lead to lineages splitting or speciating. *Au contraire.* Gradual changes can lead to reproductive isolation (and the key characteristic differentiating one species from another is the inability to interbreed). The best examples for understanding speciation involve ring species, species where some but not all subpopulations can interbreed. The geographically adjacent populations are different enough from each other such that reproductive isolation occurs in some but not all cases. Moreover, we can select for the start of reproductive isolation in the laboratory. For more on when, how, and why speciation occurs, and for a more complete explanation of ring species, go to Chapter 8.

- ✔ **Evolution of new characteristics:** Some folks insist that mutations can affect existing structures or traits but can't be responsible for new ones. Except that they can. As Chapter 15 explains, the process of gene duplication can result in multiple copies of a gene, and these copies can evolve along different trajectories. Changes in one copy that would have been deleterious in the absence of the other copy now *are not* deleterious (because you've got a spare copy) and *are* potentially advantageous. Through this process, the number of genes present in the organism can increase and diversify in function.

- ✔ **Big changes in physical characteristics:** If I'm starting to sound like a broken record, it's because the same goes for big changes in body structures: Small changes can produce big results. See Chapter 14 for more details on the evolution of development.

# No Missing Link Means No Proof

In the period immediately after Darwin published *On the Origin of Species,* there was a lot of talk about missing links. If humans and apes were relatives, where was the fossil evidence? There wasn't a good answer back then, but fast forward to today and the answer is easy: in museums all over the world!

We've found a wealth of fossils, everything from recent relatives like Neanderthals, to more distant relations whose two-legged stance puts them clearly in our part of the tree of life but whose tiny brains suggest that *we* wonder about their lives a fair bit more than they probably did. Every few months a story appears in the news about some new fossil discovery. Modern paleontologists have gotten really good at finding these things! And this in spite of the fact that

- ✔ **Fossils generally are rare and hard to find.** If, as scientists suspect, speciation occurs more often in small, isolated populations (refer to Chapter 8), finding such fossils would be much harder. But just because a certain fossil hasn't been found doesn't mean that it doesn't exist. Just that we need to keep looking — and look we do.

- ✔ **As scientists get better at fossil-hunting, they've found more and more fossils, some of which definitely qualify as transitional life forms.** Although scientists haven't found all missing links, they have found series of fossils that document transitions for many cases. Any of the following creatures discovered in the fossil record could be considered missing links, for example:

  - The fish with legs

  - The whale with legs

  - The series of feathered dinosaurs leading up to flight

Who knows what we'll see on tomorrow's news!

# It Can't Account for Everything: Enter the Intelligent Designer

There are almost as many descriptions of intelligent design (ID) as there are proponents of the theory, and some even allow a limited role for evolution via natural selection. But all the versions of ID have one thing in common: the belief that some things in the biological world could not have come about without a "designer."

Proponents of ID argue that some structures (or systems, processes, or whatever) in the biological world clearly show that they were produced by an intelligent designer. This designer may be a divine entity but doesn't need to be. Leading proponents of ID, testifying in a court of law, have suggested that the designer could be a space alien or a time-traveling biologist.

ID proponents identify complex biological structures and then state that these structures could not have been the product of natural selection and, therefore, are evidence of the designer. Yet they don't produce any testable hypotheses. Their arguments aren't scientific — regardless of the scientific terms and language they use — but theological, aliens and time travelers notwithstanding. They can't say, exactly, what it is that allows them to conclude that one structure shows the hand of the designer and another one doesn't. They just seem to know it when they see it. Many books are written on the subject of ID, but none of them share the methodology that would allow a student of ID to learn how these decisions are reached.

In this book, I don't attempt to address in detail the intricacies of religious beliefs. Religion can be a powerful force for good, but it is no more appropriate for a religious viewpoint to try to interject itself into the scientific process than it would be for the scientific view point to claim special knowledge of the mysteries of religion.

# It Can't Create Complex Structures

Irreducible complexity is the key component of most of the arguments put forth by the ID camp. Proponents of ID argue that extremely complicated structures, such as the eye, could not have evolved through a series of small steps because an eye is so complicated that it won't work without all its parts. When something can't work without all its parts, they conclude that it could not have been assembled one part at a time.

To bolster their argument, ID proponents quote Darwin himself, claiming that his very words support their argument against evolution. Well, here's what Darwin actually said about the structure of the eye, an organ he considered "of extreme perfection complication" *(Note:* The italics are mine and they highlight the part of this quote that ID proponents don't share):

> To suppose that the eye, with all its inimitable contrivances for adjusting the focus to different distances, for admitting different amounts of light, and for the correction of spherical and chromatic aberration, could have been formed by natural selection, seems, I freely confess, absurd in the highest possible degree. *Yet reason tells me that if numerous gradations from a perfect and complex eye to one very imperfect and simple, each grade being useful to its possessor, can be shown to exist; if further, the eye does vary ever so slightly, and the variations be inherited, which is certainly*

*the case; and if any variation or modification in the organ be ever useful to an animal under changing conditions of life, then the difficulty of believing that a perfect and complex eye could be formed by natural selection, though insuperable by our imagination, can hardly be considered real.*

Time and again, for pretty much whatever the ID camp claims couldn't have evolved incrementally, evolutionary biologists have identified the intermediate steps that led to the complex structure. Two classic examples of highly complex structures that evolved precisely through intermediate steps are the eye and blood-clotting factors.

Just because a system is made up of a series of parts doesn't mean that those parts evolved to perform the functions they now perform. Take, for example, bacteria that have evolved to break down polychlorinated biphenyls (PCBs), which are new to the environment. Until humans made them, these very nasty chemicals didn't exist. But some bacteria have evolved very complicated biochemical mechanisms for breaking down these compounds. As it turns out, the biochemistry that allows these bacteria to degrade PCBs is kludged together from a series of other biochemical pathways that serve other functions. Such PCB-busting biochemical mechanisms seem irreducibly complex, but the individual parts are advantageous in ways that are not related to PCB degradation.

# It Should Be Taught with ID in Science Class

ID proponents argue that the theory of evolution isn't the only theory explaining how life on Earth came to be; therefore, in the interest of fairness and balance, ID should be taught along with evolution in the science classroom.

The issue isn't what should be included in the school curriculum, but what should be included in a *science* class. This statement may sound like hair-splitting, but science instruction isn't about a simple accumulation of facts and data (even though facts and data are accumulated in the process of doing science).

Science is a way of asking questions by coming up with ideas and then trying hard to shoot them down. The ideas that scientists can't shoot down even after lots of trying become theories about how things work. And if at some point, other scientists come up with evidence that refutes these theories, they shoot the theories down.

Although the ID argument is compelling to many people, it *isn't* science. In fact, it turns the entire scientific process on its head. Instead of trying to shoot down their own premise that a designer is responsible for the complexity of the universe, ID proponents use the very complexity that they claim requires a designer to prove the existence of the designer. This reasoning is circular (and an error in logic); it's not science.

The ID argument relies on a particular world view that demands a designer. In essence, it promotes the religious viewpoint that something beyond natural processes created the world and the creatures in it. Evolution, on the other hand, is a scientific discipline; it doesn't concern itself with anything beyond what can be seen or observed in the natural world. That one deals with the supernatural and the other with the natural is the key difference between science and religion and why they don't have to be at odds.

# It's a Fringe Topic

Evolution is a central part of modern biology. In fact, making sense of most of biology concepts in the absence of an evolutionary perspective is difficult. One of the most important things that an understanding of the evolutionary process provides to the study of biology is a way to understand the effect of history.

This historical perspective is important for fields as diverse as agriculture, conservation biology, and medicine. Doctors, for example, don't worry about removing an appendix, because they have a framework in which to understand that it's a vestigial organ — that is, it may have served a purpose once, but that purpose is long gone, even though the organ isn't. And understanding how organisms evolve continues to be vital in the fight against infectious diseases (see Chapters 18 and 19 to find out why).

Conservationists seeking to save species also need to preserve biological diversity. Without genetic diversity, endangered species — even those that are making headway in the numbers game — remain vulnerable to extinction. By understanding natural variation in the evolutionary process, conservationists better understand what their conservation goals should be and how to meet them. It's probably better to save a few spotted owls from a bunch of different forests than all the spotted owls from one isolated forest, for example.

# *It's at Odds with Biblical Creation*

Quite a few people see discrepancies between the biblical creation story as they understand it and the idea of evolution. Young Earth creationism, for example, states that the Earth is only a few thousand years old and that all living organisms were created by God exactly as you see them today. Right there, you can see the areas of disagreement. This theory's creation date is at odds with most of what humans know from other fields of science — specifically, from physics and astronomy, which indicate that the Earth is about 4 billion years old. If the Earth were only a few thousand years old, the evolutionary process as scientists understand it wouldn't have had sufficient time to generate the diversity of the planet.

Old Earth creationism differs from Young Earth creationism in that it accepts that the Earth is as old as physicists and astronomers say, but it disagrees that any evolutionary processes would have occurred over that time. According to this theory, species were formed by God and did not change subsequently.

Other groups make other distinctions:

- Some allow for the possibility of small evolutionary changes that may have happened within a species over time but not for the origin of new species.

- Others allow for the possibility that speciation could have occurred within specific groups but say that larger taxonomic groups could not have arisen.

- Still others, recognizing that the Ark was only so big, have come up with a clever workaround that melds both biblical and evolutionary theory: Noah loaded the Ark with all the animals that existed on Earth at the time. Then somehow, in the few short years that followed the grounding of the Ark, the species diversified to produce the variety we see today.

In all these regards, evolution *is* at odds with a literal interpretation of biblical creation story. There's just no way around it. Many denominations of Christianity (as well as other religions), of course, have no problem with the theory of evolution or with the age of the Earth being a little over 4 billion years. Maybe, they say, that's just the way God did things.

Evolution is a fact that scientists can measure and test. As we further our understanding of the underlying processes responsible for evolution, we refine our theories about the details. If these theories ever seem at odds with particular aspects of religious belief, be assured that that was merely a consequence of following the data and never an intentional goal. The process of science has no mechanisms for addressing questions of a spiritual nature; it concerns itself solely with the natural world.

# Index

• *E* •

• *S* •

0-7645-9847-3

0-7645-2431-3

**Also available:**
- Business Plans Kit For Dummies
  0-7645-9794-9
- Economics For Dummies
  0-7645-5726-2
- Grant Writing For Dummies
  0-7645-8416-2
- Home Buying For Dummies
  0-7645-5331-3
- Managing For Dummies
  0-7645-1771-6
- Marketing For Dummies
  0-7645-5600-2

- Personal Finance For Dummies
  0-7645-2590-5*
- Resumes For Dummies
  0-7645-5471-9
- Selling For Dummies
  0-7645-5363-1
- Six Sigma For Dummies
  0-7645-6798-5
- Small Business Kit For Dummies
  0-7645-5984-2
- Starting an eBay Business For Dummies
  0-7645-6924-4
- Your Dream Career For Dummies
  0-7645-9795-7

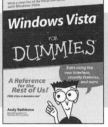

0-470-05432-8

0-471-75421-8

**Also available:**
- Cleaning Windows Vista For Dummies
  0-471-78293-9
- Excel 2007 For Dummies
  0-470-03737-7
- Mac OS X Tiger For Dummies
  0-7645-7675-5
- MacBook For Dummies
  0-470-04859-X
- Macs For Dummies
  0-470-04849-2
- Office 2007 For Dummies
  0-470-00923-3

- Outlook 2007 For Dummies
  0-470-03830-6
- PCs For Dummies
  0-7645-8958-X
- Salesforce.com For Dummies
  0-470-04893-X
- Upgrading & Fixing Laptops For Dummies
  0-7645-8959-8
- Word 2007 For Dummies
  0-470-03658-3
- Quicken 2007 For Dummies
  0-470-04600-7

0-7645-8404-9

0-7645-9904-6

**Also available:**
- Candy Making For Dummies
  0-7645-9734-5
- Card Games For Dummies
  0-7645-9910-0
- Crocheting For Dummies
  0-7645-4151-X
- Dog Training For Dummies
  0-7645-0410-9
- Healthy Carb Cookbook For Dummies
  0-7645-8476-6
- Home Maintenance For Dummies
  0-7645-5215-5

- Horses For Dummies
  0-7645-9797-3
- Jewelry Making & Beading For Dummies
  0-7645-2571-9
- Orchids For Dummies
  0-7645-6759-4
- Puppies For Dummies
  0-7645-5255-4
- Rock Guitar For Dummies
  0-7645-5356-9
- Sewing For Dummies
  0-7645-6847-7
- Singing For Dummies
  0-7645-2475-5

0-470-04529-9

0-470-04894-8

**Also available:**
- Blogging For Dummies
  0-471-77084-1
- Digital Photography For Dummies
  0-7645-9802-3
- Digital Photography All-in-One Desk Reference For Dummies
  0-470-03743-1
- Digital SLR Cameras and Photography For Dummies
  0-7645-9803-1
- eBay Business All-in-One Desk Reference For Dummies
  0-7645-8438-3
- HDTV For Dummies
  0-470-09673-X

- Home Entertainment PCs For Dummies
  0-470-05523-5
- MySpace For Dummies
  0-470-09529-6
- Search Engine Optimization For Dummies
  0-471-97998-8
- Skype For Dummies
  0-470-04891-3
- The Internet For Dummies
  0-7645-8996-2
- Wiring Your Digital Home For Dummies
  0-471-91830-X

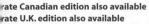
rate Canadian edition also available
rate U.K. edition also available

e wherever books are sold. For more information or to order direct: U.S. customers visit www.dummies.com or call 1-877-762-2974.
tomers visit www.wileyeurope.com or call 0800 243407. Canadian customers visit www.wiley.ca or call 1-800-567-4797.

## SPORTS, FITNESS, PARENTING, RELIGION & SPIRITUALITY

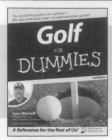

0-471-76871-5

0-7645-7841-3

**Also available:**
- Catholicism For Dummies
  0-7645-5391-7
- Exercise Balls For Dummies
  0-7645-5623-1
- Fitness For Dummies
  0-7645-7851-0
- Football For Dummies
  0-7645-3936-1
- Judaism For Dummies
  0-7645-5299-6
- Potty Training For Dummies
  0-7645-5417-4
- Buddhism For Dummies
  0-7645-5359-3

- Pregnancy For Dummies
  0-7645-4483-7 †
- Ten Minute Tone-Ups For Dummi
  0-7645-7207-5
- NASCAR For Dummies
  0-7645-7681-X
- Religion For Dummies
  0-7645-5264-3
- Soccer For Dummies
  0-7645-5229-5
- Women in the Bible For Dummie
  0-7645-8475-8

## TRAVEL

0-7645-7749-2

0-7645-6945-7

**Also available:**
- Alaska For Dummies
  0-7645-7746-8
- Cruise Vacations For Dummies
  0-7645-6941-4
- England For Dummies
  0-7645-4276-1
- Europe For Dummies
  0-7645-7529-5
- Germany For Dummies
  0-7645-7823-5
- Hawaii For Dummies
  0-7645-7402-7

- Italy For Dummies
  0-7645-7386-1
- Las Vegas For Dummies
  0-7645-7382-9
- London For Dummies
  0-7645-4277-X
- Paris For Dummies
  0-7645-7630-5
- RV Vacations For Dummies
  0-7645-4442-X
- Walt Disney World & Orlando
  For Dummies
  0-7645-9660-8

## GRAPHICS, DESIGN & WEB DEVELOPMENT

0-7645-8815-X

0-7645-9571-7

**Also available:**
- 3D Game Animation For Dummies
  0-7645-8789-7
- AutoCAD 2006 For Dummies
  0-7645-8925-3
- Building a Web Site For Dummies
  0-7645-7144-3
- Creating Web Pages For Dummies
  0-470-08030-2
- Creating Web Pages All-in-One Desk
  Reference For Dummies
  0-7645-4345-8
- Dreamweaver 8 For Dummies
  0-7645-9649-7

- InDesign CS2 For Dummies
  0-7645-9572-5
- Macromedia Flash 8 For Dummie
  0-7645-9691-8
- Photoshop CS2 and Digital
  Photography For Dummies
  0-7645-9580-6
- Photoshop Elements 4 For Dumm
  0-471-77483-9
- Syndicating Web Sites with RSS Fe
  For Dummies
  0-7645-8848-6
- Yahoo! SiteBuilder For Dummies
  0-7645-9800-7

## NETWORKING, SECURITY, PROGRAMMING & DATABASES

0-7645-7728-X

0-471-74940-0

**Also available:**
- Access 2007 For Dummies
  0-470-04612-0
- ASP.NET 2 For Dummies
  0-7645-7907-X
- C# 2005 For Dummies
  0-7645-9704-3
- Hacking For Dummies
  0-470-05235-X
- Hacking Wireless Networks
  For Dummies
  0-7645-9730-2
- Java For Dummies
  0-470-08716-1

- Microsoft SQL Server 2005 For Dum
  0-7645-7755-7
- Networking All-in-One Desk Refer
  For Dummies
  0-7645-9939-9
- Preventing Identity Theft For Dumm
  0-7645-7336-5
- Telecom For Dummies
  0-471-77085-X
- Visual Studio 2005 All-in-One Des
  Reference For Dummies
  0-7645-9775-2
- XML For Dummies
  0-7645-8845-1

# LTH & SELF-HELP

0-7645-8450-2

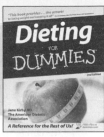

0-7645-4149-8

### Also available:

Bipolar Disorder For Dummies
0-7645-8451-0

Chemotherapy and Radiation
For Dummies
0-7645-7832-4

Controlling Cholesterol For Dummies
0-7645-5440-9

Diabetes For Dummies
0-7645-6820-5* †

Divorce For Dummies
0-7645-8417-0 †

Fibromyalgia For Dummies
0-7645-5441-7

Low-Calorie Dieting For Dummies
0-7645-9905-4

Meditation For Dummies
0-471-77774-9

Osteoporosis For Dummies
0-7645-7621-6

Overcoming Anxiety For Dummies
0-7645-5447-6

Reiki For Dummies
0-7645-9907-0

Stress Management For Dummies
0-7645-5144-2

---

# CATION, HISTORY, REFERENCE & TEST PREPARATION

0-7645-8381-6

0-7645-9554-7

### Also available:

The ACT For Dummies
0-7645-9652-7

Algebra For Dummies
0-7645-5325-9

Algebra Workbook For Dummies
0-7645-8467-7

Astronomy For Dummies
0-7645-8465-0

Calculus For Dummies
0-7645-2498-4

Chemistry For Dummies
0-7645-5430-1

Forensics For Dummies
0-7645-5580-4

Freemasons For Dummies
0-7645-9796-5

French For Dummies
0-7645-5193-0

Geometry For Dummies
0-7645-5324-0

Organic Chemistry I For Dummies
0-7645-6902-3

The SAT I For Dummies
0-7645-7193-1

Spanish For Dummies
0-7645-5194-9

Statistics For Dummies
0-7645-5423-9

---

# Get smart @ dummies.com®

- **Find a full list of Dummies titles**
- **Look into loads of FREE on-site articles**
- **Sign up for FREE eTips e-mailed to you weekly**
- **See what other products carry the Dummies name**
- **Shop directly from the Dummies bookstore**
- **Enter to win new prizes every month!**

---

* rate Canadian edition also available

* rate U.K. edition also available

ble wherever books are sold. For more information or to order direct: U.S. customers visit www.dummies.com or call 1-877-762-2974.
stomers visit www.wileyeurope.com or call 0800 243407. Canadian customers visit www.wiley.ca or call 1-800-567-4797.